RISE OF THE RED ENGINEERS

EAST-WEST CENTER

SERIES ON *Contemporary Issues in Asia and the Pacific*

A Series Sponsored by the East-West Center

CONTEMPORARY ISSUES IN ASIA AND THE PACIFIC

John T. Sidel and Geoffrey M. White, Series Co-Editors

A collaborative effort by Stanford University Press and the East-West Center, this series focuses on issues of contemporary significance in the Asia Pacific region, most notably political, social, cultural, and economic change. The series seeks books that focus on topics of regional importance, on problems that cross disciplinary boundaries, and that have the capacity to reach academic and other interested audiences.

The East-West Center is an education and research organization established by the U.S. Congress in 1960 to strengthen relations and understanding among the peoples and nations of Asia, the Pacific, and the United States. The Center contributes to a peaceful, prosperous, and just Asia Pacific community by serving as a vigorous hub for cooperative research, education, and dialogue on critical issues of common concern to the Asia Pacific region and the United States. Funding for the Center comes from the U.S. government, with additional support provided by private agencies, individuals, foundations, corporations, and the governments of the region.

JOEL ANDREAS

Rise of the Red Engineers

The Cultural Revolution and the Origins of China's New Class

Stanford University Press · *Stanford, California*

Stanford University Press
Stanford, California

Printed in the United States of America on acid-free, archival-quality paper
Library of Congress Cataloging-in-Publication Data
Andreas, Joel.
Rise of the red engineers : the Cultural Revolution and the origins of China's new class / Joel Andreas.
p. cm. -- (Contemporary issues in Asia and the Pacific)
Includes bibliographical references and index.
ISBN 978-0-8047-6077-5 (cloth : alk. paper) -- ISBN 978-0-8047-6078-2 (pbk. : alk. paper)
1. Elite (Social sciences)--China--History--20th century. 2. Social classes--China--History--20th century. 3. Power (Social sciences)--China--History--20th century. 4. Engineers--China--History--20th century. 5. China--History--Cultural Revolution, 1966–1976. I. Title. II. Series: Contemporary issues in Asia and the Pacific.
HN740.Z9E413 2009
305.5'24095109045--dc22
2008043089

Typeset at Stanford University Press in 9.75/13.5 Janson Text

For my mother, Carol Andreas (1933–2004), who made me care about social injustice and set me on the path of exploring its causes.

Contents

Preface

When I first visited Tsinghua University in Beijing in 1997, my aim was to learn about the battles that took place there three decades earlier during the Cultural Revolution. I had heard about the "hundred day war" between student factions at Tsinghua and knew that one side was led by Kuai Dafu, a student whose name had become synonymous with the rebellious spirit of the period. My curiosity about the Cultural Revolution was inspired by a larger interest in the transformation of China's class structure since the 1949 Revolution, but it only gradually occurred to me, as I interviewed Tsinghua employees and alumni, that in addition to being an important site of Cultural Revolution battles, the university had for decades been at the epicenter of conflicts surrounding the emergence of a new class of technocratic officials.

Before the Cultural Revolution, Tsinghua—as China's leading school of engineering and technology—had been charged with training "Red engineers." Technocratic visions flourished at the university and students believed they would lead the country's transformation into an industrialized socialist republic. These visions were always controversial. They were at odds with the Chinese Communist Party's programmatic commitment to eliminate class distinctions, including those based on the differences between mental and manual labor, and they were foreign to most of the party's cadres, who were peasant revolutionaries who celebrated traditions born of rural warfare and harbored a deep distrust of the educated elite. Simmering tensions came to a head during the Cultural Revolution. Tsinghua became a prominent target and after factional fighting was suppressed, Mao Zedong dispatched a team of workers and soldiers to the university and charged them with eliminating elitist educational practices and preventing the school from becoming an incubator of a "bureaucratic class." For nearly a decade, the campus served as a celebrated site for implementing radical experiments in education and governance. Then, after Mao died and the radical policies of the Cultural Revolution were renounced, the university emerged as the premier training ground for the type of technically competent and politically reliable cadres required by a new regime committed to a technocratic vision of China's future. Tsinghua graduates climbed quickly to the top of party and state hierarchies.

Most of my interviewees had spent decades at the university, first as students, and then as professors or administrators, and they had experienced firsthand much of this tumultuous history. As I came to see that the university had been at the center of all of the key conflicts that ultimately produced a technocratic order, I realized the value of closely examining the history of an institution on the ground level, where policies were implemented and social relations transformed. That is how my investigation into the origins of China's technocratic class became centered on Tsinghua University. I have asked the university to bear quite a burden, as ultimately I was interested not only in the origins of the technocratic officials who rule China today, but also in the fate of the twentieth-century Communist experiment. I have tried in this account of China's premier engineering school to capture crucial elements of the trajectory of communism in power, which began as a radical social leveling project and ended as an immense bureaucratic enterprise with an elaborate social hierarchy.

I was fortunate that Tsinghua was a relatively open place. Campus apartment blocks were filled with individuals—working and retired—who were willing to share their experiences and perspectives with me, and campus libraries were filled with materials that provided a rich documentary history of the institution. Altogether I spent twenty months at the university, mostly between 1997 and 2001, gathering data from both retrospective and contemporary sources. The most important retrospective sources were interviews with nearly one hundred people, including graduates, teachers, staff, workers, and administrators (a list of interviewees is included in Appendix 2); I sought individuals who came from a variety of social origins and had different political perspectives. Retrospective published sources included memoirs, biographical sketches of prominent professors and school leaders, a university gazetteer compiling historical data, and official and semiofficial school histories. I also consulted scholarly books and articles about the university's history. Contemporary sources included official university newspapers and magazines, newspapers and pamphlets published by contending student factions, articles about the university that appeared in national newspapers and journals, as well as statistical yearbooks and administrative and political reports produced by the university. Documents were obtained from libraries in the United States and China (including libraries at Tsinghua University and its attached middle school), the Tsinghua University archives, used-book markets, and the personal collections of interviewees and others.

All accounts—retrospective and contemporary, oral and written (including the statistics recorded in official documents)—reflect the biases of the producers and the times. Contemporary and retrospective sources have complementary strengths and weaknesses. Public expression in China has been and continues to be constrained by the prevailing ideology of the period and by political considerations. During the first decades of communist power, unorchestrated political debate was possible only during two brief moments, the 1957 Party Rectification campaign and the early years of the Cultural Revolution, and even then although debates were sharply contentious they generally remained within narrow political boundaries. On the other hand, these materials recorded the political discourse of the period and interpreted events from a period perspective. Data produced retrospectively—such as interviews, memoirs, and histories—are freed from past constraints and incentives, but are subject to new sets of constraints and incentives. Moreover, individuals' recollections after several decades of profound social change have to be treated with caution. Memories of past events, motivations, and ideas not only fade with time, but they also undergo a conscious or unconscious metamorphosis as subsequent events and political and ideological changes (official, collective, and personal) make their imprint. I have, therefore, carefully compared a wide variety of sources to reconstruct historical events, including the contentious viewpoints that animated them. One advantage of focusing on a single school is that it was possible to get many different perspectives on the same events.

I have chosen to use "Tsinghua," rather than the standard *pinyin* transliteration, "Qinghua," because the former is the official preference of the university today and is therefore ubiquitous in foreign-language references to the school. The decision to use an older transliteration reflects the university's effort to recover and celebrate its prerevolutionary past, one manifestation of the reconciliation of new and old elites narrated in this book. Based on pronunciation, English speakers unfamiliar with *pinyin* and other conventional romanization systems might choose to spell the name "Chinghua."

This book has been more than a decade in the making and many more people have helped me over the years than I can acknowledge here. I am particularly grateful to all the people who took the time to tell me their stories. Many generously spent hours—even days—detailing their understandings of events, and some searched for photographs and documents to illustrate their points. (With the exception of Kuai Dafu, I have not used their real names in this book.) The people at Tsinghua University's Education Research Institute provided a congenial base for my studies, as well as a regular source of information and dialogue. I particularly want to thank Wang Xiaoyang and Wang Sunyu, who were my hosts while I was at the institute. Librarians and archivists at the institute and elsewhere at the university provided great assistance locating materials.

Tang Shaojie, who has spent many years studying the Cultural Revolution at Tsinghua, first encouraged me to make the university a research site, and I have learned a great deal from his research and from our conversations over the years. I greatly benefitted from the help and wisdom of three people in particular, Dai Jianzhong, Wu Ciaxia, and Xu Hailiang, who participated in the Cultural Revolution and have spent the years since investigating and reconsidering the events of that era in light of subsequent events. They generously provided me with many documents, and each played an important role in shaping my understanding (although none will completely agree with my interpretations). Michael Schoenhals, Richard Siao, and Wang Youqing also provided important documents that I otherwise would never have encountered.

My approach to studying the development of a dominant class in post-revolutionary China was shaped especially by mentors and colleagues at UCLA. I learned a tremendous amount from Rogers Brubaker and Michael Mann, who provided masterful examples of how to study history from a sociological perspective and continually offered insightful advice. Ivan Szelenyi's influence is especially obvious, as he steered me toward the conceptual framework of multiple types of capital and contending elites, and his work provided an exceptionally intelligent foil for my own. Rebecca Emigh, Philip Huang, William Roy, Gi-Wook Shin, Donald Treiman, and Yan Yunxiang all provided valuable counsel. Perry Anderson, Robert Brenner, Susanne Chan, Eileen Cheng, Steven Day, Clayton Dube, Jon Fox, Margaret Kuo, Mara Loveman, Mark Lupher, Meng Yue, Dylan Riley, Song Shige, Elizabeth VanderVen, Wang Chaohua, Wu Shengqing, and Wu Xiaogang read drafts of chapters or papers and gave me helpful advice. I particularly appreciate the help of Cong

Xiaoping, who in addition to reading chapters, helped analyze documentary materials and provided insights from her own experience.

Over the years, many other people have offered valuable comments, criticisms, and suggestions on portions of this book in its various stages, among them Cui Zhiyuan, Arif Dirlik, Fred Engst, Joseph Esherick, Sheila Fitzpatrick, Han Dongping, Andrew Kipnis, Li Lulu, Lu Aiguo, William Parish, Peng Yusheng, Suzanne Pepper, Stephen Philion, Paul Pickowicz, Dorothy Solinger, Su Yang, Saul Thomas, Tian Liwei, Jonathan Unger, Andrew Walder, Lynn White, Wu Lili, Wu Yiching, Yan Hairong, Yin Hongbiao, Zhan Shaohua, Zhao Dingxin, and Zheng Xiaowei.

Since I have been in Baltimore, my colleagues at Johns Hopkins, including Rina Agarwala, Giovanni Arrighi, Marta Hanson, Melvin Kohn, Tobie Meyer-Fong, William Rowe, Beverly Silver, and Kellee Tsai, have helped me bring this book to a conclusion, offering practical advice and encouragement. I owe a special debt to Kellee for her critical support and suggestions. For research and editing assistance, I thank Laila Bushra, Angela Huang, Huang Lingli, Li Meng, Li Yuyu, Sun Haitao, Wang Yingyao, and Yue Yin.

I was able to spend many months in China due to research fellowships provided by the Fulbright-Hays program of the U.S. Department of Education, Peking University, UCLA, and the Social Science Research Council. Support for writing was provided by UCLA and the Spencer Foundation.

Members of the editorial board of the Contemporary Issues in Asia and the Pacific series, and anonymous reviewers, helped shape the final form and content of this book by urging me to strengthen a number of weak points and add a more elaborate analysis of recent developments. Elisa Johnston at the East-West Center and Stacy Wagner at Stanford University Press have done a wonderful job in shepherding this book from initial manuscript through page proofs.

I am grateful to Springer Media, the Contemporary China Centre at the Australian National University, and the American Sociological Association for allowing me to use material that originally appeared in my articles, "Battling over Political and Cultural Power During the Chinese Cultural Revolution," *Theory and Society* 31 (2002), "Institutionalized Rebellion: Governing Tsinghua University During the Late Years of the Chinese Cultural Revolution," *The China Journal* 55 (2006), and "The Structure of Charismatic Mobilization: A Case Study of Rebellion During the Chinese Cultural Revolution," *American Sociological Review* 72 (2007).

Finally, I want to thank Peter Andreas, who has been both a steadfast little brother and my principal mentor in the academic world, and my wife, Ay Vinijkul, who has been my companion and most important supporter through the long process of completing this book (and who will be as happy as I am that it is finally finished).

Joel Andreas
Baltimore, Maryland
April 2008

RISE OF THE RED ENGINEERS

Introduction

China today is ruled by Red engineers. This term, which dates from the 1950s when China was embarking on Communist-style industrialization, was condemned during the Cultural Revolution and has not been revived since. In the 1980s, however, the Red engineers who received academic and political training at elite technical universities in the 1950s and early 1960s began moving into positions of power. They systematically replaced the first generation of Communist cadres, initially at the lower and middle levels, and then during the 1990s at the very highest levels of the Chinese Communist Party (CCP). Today, eight out of nine members of the Standing Committee of the party's Political Bureau were trained as engineers.

China's Red engineers, not by coincidence, resemble the officials who staffed the upper levels of the state machinery in the Soviet Union and Eastern Europe prior to 1989. The Soviet Union provided a model for China and other countries where Communist parties came to power, and for decades it was led by men who had degrees in engineering or agronomy, including Leonid Brezhnev, Alexei Kosygin, Andrei Gromyko, Yuri Andropov, and Mikhail Gorbachev. Red engineers in the Soviet Union, China, and elsewhere ruled socialist societies that in many ways resembled the technocratic vision of Henri Saint-Simon. In the early nineteenth century, Saint-Simon's followers had envisioned an industrial order that would transcend the avarice of capitalism by converting the means of production into public property and conducting economic planning based on scientific principles. Although they saw inheritance and private property as unjust and inimical to progress, theirs was a supremely elitist vision, in which a talented and enlightened group of industrial leaders, scientists, and engineers would govern society.[1]

Of course, the Chinese Communist Party, like the Russian Bolsheviks, had originally championed a Marxist rather than a Saint-Simonian vision of socialism. Marx adopted the basic premises of Saint-Simonian socialism, but rejected its hierarchical character. While Saint-Simon endeavored to establish a society ruled by the talented, Marx sought to eliminate all class distinctions, including the distinction between mental and manual labor, and while Saint-Simon set out to organize a movement of the educated elite and recruited an enthusiastic following among graduates of Paris's prestigious École Polytechnique, Marx called on the proletariat to serve as the revolutionary vanguard because, he reasoned, they had little to lose by doing away with the existing class hierarchy.[2] It was Marx's ideas, with their egalitarian thrust, rather than those of Saint-Simon, which eventually galvanized the momentous socialist movements of the nineteenth and twentieth centuries. Moreover, the Communist parties that took power in Russia, China, and other countries espoused a particularly radical interpretation of Marx's ideas that committed them to mobilizing the most downtrodden classes, seizing state power by force, crushing the resistance of the old elites, and ruthlessly eliminating all class distinctions. After they came to power, these parties did, indeed, radically change the class order, but in the end they did not do away with class distinctions. The class hierarchy based on private property was destroyed, but a new hierarchy based on political and cultural power emerged, with a class of party technocrats on top.

In this book, I seek to explain how and why the Chinese Communists ultimately replaced Marx's vision of a classless society with one worthy of Saint-Simon. One possible explanation is that victorious Communist parties, despite their class-leveling rhetoric, always intended to build a technocratic society. This view has been put forward most cogently by George Konrad and Ivan Szelenyi in their seminal book, *Intellectuals on the Road to Class Power*. Konrad and Szelenyi argued that Communist parties, their claims to represent the proletariat notwithstanding, were actually the vanguard of the intelligentsia. The Communists' fundamental goal, they wrote, was to fulfill intellectuals' long-held ambition to displace aristocrats and capitalists and take the reins of society into their own hands by substituting public for private property and planning for the market. Konrad and Szelenyi were among a wave of scholars in the 1960s and 1970s who emphasized the technocratic nature of Soviet and Eastern European elites, and they produced a brilliant insider's account of how power was distributed and wielded in Soviet-type societies.[3] Although they were personally involved in the "New Class project" they described, they subjected it to an unsentimental analysis of interests, revealing the connections between class power and the celebration of knowledge and science. The tension in their narrative is provided by contention between the Communist vanguard, who stubbornly tried to maintain the prerogatives of political power, and the wider intellectual class, who sought to make knowledge the principal basis of class power. This struggle revolved around the competing claims of political capital (party membership and political connections) and cultural capital (knowledge and academic credentials). Konrad and Szelenyi predicted that the rational premises of socialist planning would ultimately lead to the triumph of cultural capital, fulfilling what they claimed was the true essence of the Communist mission.[4] Their argument was provocative and compelling, and it profoundly influenced scholarly discussion about the class structure of socialist societies.

Although the technocratic characteristics of China's New Class fit Konrad and Szelenyi's theory to a tee, its history does not. One dramatically incongruous element in this history was the harsh attacks on intellectuals during the first decades of Communist power in China. For over a quarter century, the Chinese Communist Party worked tenaciously to eliminate the class distinctions that separated intellectuals from workers and peasants. In their most radical moments, the Chinese Communists systematically discriminated against members of the old educated classes, eliminated entrance examinations and filled university classrooms with villagers who had not attended high school, denigrated the value of abstract knowledge, sent intellectuals to live in villages to be reeducated by peasants, and strived to level educational differences by providing nine or ten years of education to everyone—children of intellectuals, workers, and peasants alike—and then sending them to work. These hardly seem like policies invented by champions of the intelligentsia. Moreover, China's program of cultural leveling was not unique. The Bolsheviks made Marx's goal of eliminating the distinction between mental and manual labor into a central tenet of their program, and in the early years of Soviet power—especially during the period of cultural radicalism that accompanied the First Five Year Plan (1928–32)—they pursued radical education policies similar to those later implemented in China. Konrad and Szelenyi's brief explanation for early Communist policies hostile to the educated elite—that they were part of a "costly but indispensable detour" necessary to build the strong state technocracy required—do not seem adequate.[5] It is even more difficult to fit into Konrad and Szelenyi's theory Mao Zedong's efforts to undermine the bureaucratic power of party officials. These efforts included harsh campaigns against cadres' privileges and abuse of power, which reached a crescendo during the Chinese Cultural Revolution, when Mao called on workers, peasants, and students to overthrow local party authorities in order to prevent Communist officials from becoming a "bureaucratic class."

In this book, I tell the story of the rise of China's New Class. In order to explain the incongruities noted above, this account diverges from Konrad and Szelenyi's theory in two ways. First, I do not insist that Communist cadres, most of whom were peasant revolutionaries with relatively little education, play the role of vanguard of the intelligentsia. Instead, I develop an analytic framework that I believe more accurately describes much of the conflict that followed the 1949 Revolution: contention and cooperation between a new political elite and an old educated elite. The new political elite was largely composed of peasant revolutionaries and the old educated elite was mainly made up of members of the dispossessed propertied classes. Although there was overlap between the two groups, for the most part their social origins were distinct, they had discordant value systems, and they relied on different types of resources. Members of the first group controlled the reins of political power but had little education, while members of the second group faced severe political handicaps but possessed substantial cultural resources. The New Class in China, I argue, was the product of a violent and contentious process that ultimately culminated in the convergence of these old and new elites.

Second, I take seriously Communist efforts to eliminate class distinctions. While Konrad and Szelenyi treated the emergence of a technocratic class as the achievement of Communist class-building strategy, I treat it as the failure of Communist class-leveling efforts. If the New Class was simply the product of deliberate construction, the process need not have taken so many harrowing twists and turns. I will argue that Communist parties fundamentally changed course, abandoning the road of class leveling and taking instead a technocratic road. They were converted from enemies into champions of cultural capital, a transformation that Konrad and Szelenyi obscured in their endeavor to portray the Communist movement from its inception as a technocratic project of the intelligentsia. Thus, I will argue that Konrad and Szelenyi's boldest claim—that victorious Communist revolutionaries *deliberately* built a technocratic order ruled by an educated elite—does not hold true. If we remove this supposition of intention, however, it becomes possible to ask a more interesting question: Why—despite forceful efforts to the contrary—did the Communist project result in the creation of a new dominant class of Red experts? Answering this question is the chief purpose of this book.

There are powerful reasons to conclude that this result was inevitable. Every Leninist state that survived for a significant length of time eventually gave rise to a technocratic class order, and the technical requirements of economic planning provide a ready functional explanation for the consistency of this outcome. Some, therefore, might be inclined to close the case without further investigation. It is always wise, however, to examine cautiously claims of historical inevitability and functional necessity, especially when one is investigating the origins of a system of social differentiation in which group interests are involved. Even though I am not convinced by Konrad and Szelenyi's account, I share with them an inclination to explain history as the product of conflicts between interested parties. Moreover, we can learn a great deal by studying the problems encountered by the Communist class-leveling projects of the twentieth century. Marxist revolutionaries vowed not only to eliminate private wealth, but also to redistribute political and cultural power to the masses, and their radical democratic and egalitarian rhetoric was converted into far-reaching social experiments. By carefully scrutinizing these experiments and identifying the reasons they failed, we can inform future redistribution efforts, which will undoubtedly run into some of the same problems.

Research Strategy

I chose to investigate China because it is an extreme case. China had much in common with other states that implemented the Soviet model; what makes the Chinese case stand out is the Cultural Revolution. As the following chapters will show, the Cultural Revolution was a determined effort to undermine the political and cultural foundations of an emergent stratum of Red experts. Scholars investigating other Leninist states have appropriately compared political campaigns and policies in these countries to aspects of the Chinese

Cultural Revolution, but no other state experienced such a protracted, tenacious, and disruptive effort to prevent the emergence of a technocratic class. If we want to know whether the rise of such a class was inevitable, it makes sense to study China.

This book is based on a case study of a single educational institution, Tsinghua University in Beijing. Tsinghua is China's consummate trainer of Red engineers. It is the country's premier engineering school, and the university's party organization is renowned for grooming political cadres. Today, Tsinghua graduates occupy key positions in the upper echelons of the party and state bureaucracies, and one-third of the members of the Political Bureau's Standing Committee, including Secretary General Hu Jintao, are alumni.[6]

China's Red engineers have been cultivated by two highly selective credentialing systems, one academic and the other political, both of which were modeled after Soviet institutions. The academic credentialing system consists of a pyramid of increasingly selective schools, starting with primary schools and culminating in a small number of elite universities. The political credentialing system is the party's recruitment apparatus, consisting of a parallel hierarchy of increasingly selective organizations. In primary school, young people compete to join the Young Pioneers, and in secondary and tertiary school they compete to join the Communist Youth League and then the Communist Party. Because of both technical requirements and ideological inclinations connected with industrialization, Tsinghua and other elite engineering schools are located at the pinnacle of both credentialing systems.

I chose to study a university because I wanted to be able to closely examine the struggles surrounding the academic and political credentialing systems, and I selected Tsinghua because it was a singularly important battlefield. Whether policy veered to the Left or the Right, the university served as a model for other schools to follow. During the decades after 1949, Tsinghua grew into a sprawling multifaceted institution that encompassed elite primary and secondary schools, numerous factories, onsite programs to train workers, peasants, and "worker-peasant cadres," and satellite schools in remote work sites and villages. All of these programs served as showcases for highly contentious social experiments. Conducting a detailed case study allowed me to analyze, from a ground-level perspective, how both the academic and political credentialing systems actually functioned, how they changed, and how the conflicts over them unfolded. I was able to observe how radically different education policies were implemented, and follow the construction of the party and Youth League organizations at the university, as well as their collapse during the Cultural Revolution and their subsequent reorganization. By closely following changes in a particular institution, I was able to develop a much richer and more concrete story than if I had simply studied conflicts among top party leaders, the evolution of national policy, and countrywide statistical trends.

In this book, I am attempting—in the words of Michael Burawoy—to "extract the general from the unique."[7] Tsinghua is hardly a typical Chinese university; it is located at the apex of the education system, and other schools never had the resources—and often did not have the inclination—to fully implement the exemplary policies and programs developed at Tsinghua. In the following chapters, I will often point out ways in which Tsinghua was peculiar or unique. Nevertheless, the battles at the university were emblematic of wider conflicts, and we can learn much about these conflicts by examining how they played out at Tsinghua, which was always at the epicenter. China was also hardly typical of Leninist states. Countries that implemented versions of the Soviet model have so much in common, however, that it is worthwhile developing a common theoretical framework. Students of the early history of the Soviet Union and other countries in which Communist parties came to power by means of indigenous revolutions will surely recognize a family resemblance in many of the contradictions, conflicts, and policies described in the following pages. After carefully analyzing the Chinese case, with all of its irreducible peculiarities, it will be possible to compare cases, and draw more general conclusions.

PREVIOUS SCHOLARSHIP

The territory covered in this book has already been partially charted by others. Four genres of scholarly literature, in particular, extensively overlap my efforts. Central elements of this book—two elite groups and two credentialing systems—each figure prominently in one of these four genres. The first two are concerned with Communist cadres and intellectuals, while the third and fourth examine China's education and political systems. Scholars writing in the first genre have recounted how a party of poorly educated peasant revolutionaries was transformed into a party of technocratic officials.[8] Although most Communist cadres received at least a modicum of technical training after 1949, the basic story told in this genre of scholarship is of a generational change that takes place after an epochal decision in 1978 to emphasize technical over political qualifications. Remnants of the old elite classes remain in the background and the class origins of the new technocratic elite receive little attention. Scholars writing in the second genre have recounted the troubled relationship between Chinese intellectuals and the Communist regime.[9] During the Mao era, this is mainly a story of conflict, in which the Anti-Rightist movement and the Cultural Revolution loom large. Intellectuals are employed by the party/state, but are also persecuted by it; they are offered opportunities to serve, but only on Communist terms, and they must choose whether to collaborate and try to influence policy or resist. Some intellectuals join the party, and during the reform era conditions improve greatly, but the protagonists remain unchanged: intellectuals on one side and the party/state on the other. Although conflict is central to these accounts, group interests scarcely register; instead, the battles are typically about protecting space for scientific endeavor, intellectual autonomy, and humanistic ideals against the imposition of party/state dictates.

Scholarship on the Chinese education system has analyzed radical changes in education policy, while scholarship on the political system has investigated the evolution of the party system, including its breakdown during the Cultural Revolution.[10] Some scholars have explored the ways in which each of these systems has served as a mechanism of class differentiation, but the main analytical interest of most work on the education system has been the efficacy of policies in terms of conventional educational goals (quality and quantity of training), and the main analytical interest of most work on the political system has been the efficacy of policies in terms of conventional political goals (political and social control).

While the present account has benefited greatly from insights derived from scholars writing in all four genres, my research agenda is different, and it has led me to tell a story that has not been told by the works in any of these genres. Although the first two genres illuminate much about the trajectories of one or the other of the two elite groups at the center of this book, neither captures the contentious process of inter-elite convergence described in the following pages. The third and fourth genres tell us a great deal about each of the two credentialing systems at the center of this book, but they largely miss the interaction between them. Works on the education system mention the political system in passing, and works on the political system mention the education system in passing, but their analytical interests are typically confined to one realm or the other. This book is about both systems, and its analytical interest in each is the same: how the system reproduced class differentiation. Moreover, I am particularly concerned with the links between the two systems, and the political struggles chronicled in this book almost always involved both systems and both elite groups.

A fifth genre, with a narrower scope, has analyzed the social bases of contending local factions during the Cultural Revolution.[11] These accounts highlight conflict between intellectuals and party officials, and between children of the two groups. Moreover, they identify educational and political admissions policies—some of which benefited children of intellectuals, while others benefited children of party officials—as key objects of contention and determinants of factional alignment. Thus, a central theme of the present account—inter-elite conflict over academic and political credentialing policies—fits well into this genre, and my analysis of student factional struggles at Tsinghua during the Cultural Revolution (see Chapters 4 and 5) will engage these accounts in detail. While they stressed inter-elite conflict, however, I highlight strong manifestations of inter-elite unity even at the height of the battles of the Cultural Revolution, and I present these battles as part of a longer process of inter-elite convergence.

CONCEPTUAL FRAMEWORK

I use the conceptual framework Pierre Bourdieu developed to analyze class distinctions based on different types of capital.[12] Although Bourdieu mainly employed his tripartite framework—economic, cultural, and social capital—to analyze the class structure of a stable capitalist society, others have found it just as useful for analyzing radical changes in class structure. Szelenyi and others have been able to cogently describe the transitions to and from socialism in terms of changes in the relative importance of these three types of capital.[13] Communist regimes eliminated economic capital by abolishing private property in the means of production, and although control over these means of production still mattered, access to control was no longer provided by private ownership, but rather by cultural and social capital. Because Eastern European societies were dominated by Communist parties, the key form of social capital was political. As a result, class position was largely determined by an individual's stock of cultural and political capital.[14]

When Bourdieu spoke of cultural capital, the assets he had in mind—educational credentials and knowledge that provide access to advantageous class positions—are largely the same as those that many economists and sociologists discuss under the rubric of human capital.[15] The two terms, however, signal different analytic interests. While theorists of human capital investigate how returns on investment facilitate individual and social progress, Bourdieu investigated how individuals and groups use the institutions that underpin cultural capital to reproduce class privilege and power. Political capital, in Bourdieu's framework, is also about privilege and power. He conceived of political capital as a form of social capital, which he defined as "the actual or potential resources which are linked to the possession of a durable network of more or less institutionalized relationships of mutual acquaintance and recognition—in other words, to membership in a group—which provides each of its members with the backing of a collectively-owned capital, a 'credential' which entitles them to credit."[16] In the countries of the Soviet bloc, Bourdieu agreed, the most important form of social capital was political, based on party membership.[17]

Because I have chosen to focus on conflicts surrounding academic and political credentialing systems, I am mainly concerned with what Bourdieu called *institutionalized* forms of capital. The credentials distributed by these systems—including academic certificates and party membership—are critical resources used to gain access to advantageous class positions. These institutionalized forms of political and cultural capital, however, are closely connected to less tangible *embodied* forms. In the cultural field, embodied capital consists of the actual knowledge possessed by an individual, including knowledge imparted as part of the school curricula as well as a broader range of cultural competences—such as manners and tastes—that distinguish the educated from the uneducated classes. Embodied political capital consists of the social networks an individual cultivates around a party organization. In this book, unfortunately, I have been compelled to give embodied forms of capital less attention than they deserve in order to focus on the institutionalized forms.

This book is essentially an analysis of struggles over the redistribution of different types of capital. Like Szelenyi, I am particularly interested in the ability of elites to maintain their social positions across revolutionary social transformations that undermined the value of one type of capital, while enhancing the value of others. In addition to the fate of elite groups, however, I am interested in the results of Communist redistribution schemes intended to eliminate class distinctions by dispersing the possession of capital. Bourdieu's tripartite framework provides a useful conceptual template to gauge the results of Communist class-leveling efforts. The extent of class differentiation can be appraised in terms of the concentration of capital in the economic, cultural, and political fields. Any resource—whether physical property, knowledge, or political power—can only serve as a means of class differentiation to the extent that it is distributed unequally. Class power is based on the concentration of these resources in the hands of a minority and it is perpetuated by institutions that reproduce this unequal distribution. Redistribution can transfer capital from one elite group to another, further concentrate capital in the hands of an elite group, or disperse capital more widely through the population. Policies that further concentrate capital increase the gap between classes, while policies that disperse capital diminish this gap.

Tumultuous Rise of China's New Class

Stripped to its most basic elements, the story of the rise of China's New Class, as told in this book, can be summarized as follows: In the first years after the 1949 Chinese Revolution, economic and cultural capital were concentrated in the hands of the old elite classes, while political capital was concentrated in the hands of the new Communist elite, made up largely of peasant revolutionaries. The new regime first redistributed economic capital, dispossessing the old elites and converting the means of production into state and collective property. Although ownership was nominally public, control was concentrated in state and collective offices, and access to these offices was determined largely by possession of political and cultural capital. Having virtually eliminated economic capital, the CCP turned its attention to redistributing cultural capital, with the intention of further undermining the advantages of the old elite, an endeavor that reached its most radical point during the Cultural Revolution. The principal target of the Cultural Revolution, however, was the concentration of political capital in the hands of the new Communist elite. At Mao's instigation, grassroots insurgents challenged the power of local party officials, precipitating two years of factional violence. The upheaval initially exacerbated tensions between the old and new elites, but Mao's simultaneous attacks on both groups ultimately forged inter-elite unity. After Mao's death in 1976, the new CCP leadership renounced class leveling and reconciled with the old elite. This facilitated the consolidation of a technocratic class order and the emergence of a New Class, which had roots in both the old and new elites and combined their political and cultural assets.

Tsinghua University is a narrow frame through which to tell this story, one that misses the top echelons of power and the economic infrastructure of the country. The university, however, provides an excellent vantage point to closely observe the evolution of the academic and political credentialing systems, which became the key institutional foundations of the technocratic order. Tsinghua and other elite universities were at the center of the contentious convergence of old and new elites. They were an important site where the two groups met, initially in the persons of incumbent faculty and Communist cadres dispatched to take charge of the schools. Members of the first group were virtually all from well-to-do families and they had been educated in the best schools in China, the United States, Japan, and Europe, while members of the second group were battle-hardened revolutionaries who had been trained through years of rural warfare. More important, these universities were selecting and training students who would become the Red and expert elite. Children from both elite groups, along with a small but growing number of children from working-class and peasant families, filled the classrooms of top-ranked universities, where they competed not only academically, but also to join the Youth League and the party, striving to become both Red and expert.

Tsinghua became *the* focal point of conflict over both the academic and political credentialing systems. During the first seventeen years of Communist power, the university cultivated a reputation as the "cradle of Red engineers," and for this reason Tsinghua and its leaders became prominent targets during the Cultural Revolution. The university served as the base of the most influential of the contending student factions, and after the suppression of a freewheeling factional struggle it was taken over by radical leaders determined to eliminate educational elitism and undermine the bureaucratic power of the party officialdom. Then, after the Cultural Revolution was repudiated following Mao's death, the university cemented its position as the premier training ground for the type of technocratic cadres preferred by the party's new leadership. Tsinghua, therefore, provided an ideal site to analyze in detail the contentious process through which old and new elites coalesced into a New Class and Mao's failed efforts to block the way.

This book is composed of four chronological parts, each of which is divided into thematic chapters. The first part, covering the period between the Communist seizure of power in 1949 and the outbreak of the Cultural Revolution in 1966, is composed of three chapters. The first two chapters examine the controversies and conflicts that surrounded the construction of the political and academic credentialing systems at Tsinghua. Chapter 3 recounts how the CCP's determination to combine Redness and expertise, originally driven by class-war logic, ultimately promoted the coalescence of the new and old elites. The chapter closes by identifying powerful political and structural obstacles that effectively blocked the rise of a technocratic class.

The second part recounts the early years of the Cultural Revolution, from 1966 to 1968, when Mao called on students, workers, and peasants to attack both the political and educated elites. Through a detailed examination of how the ensuing factional struggle unfolded at Tsinghua University and its attached middle school, Chapters 4 and 5 show that while the movement initially greatly aggravated conflict between the old and new elites, giving rise to an explosion of factional fighting between children of Communist cadres and children of intellectuals, ultimately it also forged inter-elite unity.

The third part covers the late years of the Cultural Revolution, from 1968 to 1976, when Mao attempted to institutionalize the radical class-leveling program that had led him to launch the movement, with the explicit aim of preventing the development of a new privileged class. Chapter 6 examines the experimental system of governance created during this period, which was based on a volatile arrangement of institutionalized factional contention. Chapter 7 scrutinizes radical education policies designed to eliminate occupational distinctions between mental and manual labor, and Chapter 8 looks into the system of "mass recommendation" created to replace college entrance examinations, which was designed to increase the number of students from working-class and peasant families. These chapters examine how these policies were carried out in practice and assess their potential to actually diminish class distinctions.

The fourth part examines the establishment of a technocratic order after Mao's death in 1976 and its subsequent evolution. Chapter 9 recounts how the academic and political credentialing systems were rebuilt, enhancing their capacity to select and groom a stream of Red and expert cadres to staff the upper levels of the state bureaucracy. Chapter 10 traces the convergence of old and new elites and the consolidation of a new technocratic class. Chapter 11 considers the consequences of the sweeping economic reforms that began in the 1990s, which have left China's Red engineers presiding over a peculiar variety of capitalism. The main concern of the chapter is to ascertain how China's technocratic class order, which had been based mainly on political and cultural capital, is being transformed by the reemergence of economic capital.

The concluding chapter compares the two most important twentieth-century Communist experiments, in the Soviet Union and China, and proposes revisions to New Class theory. I show that in their early years the Communist regimes in both countries forcefully pursued class leveling, but that they later decisively abandoned this goal in order to implement technocratic policies, leading to very similar results. I suggest reasons why the Chinese made this fundamental switch, and then consider whether or not it was inevitable.

PART ONE

Building Socialism (1949–1966)

Chapter One

Political Foundations of Class Power

On December 15, 1948, Communist troops advancing on Beijing arrived at the vicinity of Tsinghua University in the northeastern outskirts of the city. The day before, the university's president, Mei Yiqi, had hastily left the campus and headed south to Nanjing, which for the moment remained in the hands of the retreating Nationalist government. Almost everyone else at the university, however, decided to stay. Most members of the Tsinghua faculty were from families of more than modest wealth and social standing, and many were certainly apprehensive about the Communists' declared aim of radically redistributing wealth and alarmed by their brutal reputation. After years of war, however, many were also hopeful that a new regime would bring order and more honest government, and some were even sympathetic with the Communists' promises of creating a more egalitarian society. To whatever extent they sympathized with the Communist Party, they knew they would have to accommodate themselves to the new order it would establish. The Communists were headed toward a decisive victory in the civil war, they had broad popular support, and they brooked no opposition.

On December 17, a delegation of Communist soldiers came to the Tsinghua campus to meet with representatives of the faculty, staff, and students; the purpose of the meeting was to make arrangements to safeguard university facilities. On one side of the meeting were erudite, well-groomed intellectuals; on the other side were battle-hardened peasant revolutionaries. The Communist delegation was headed by Liu Daosheng, a peasant from Hunan Province who had joined the Communist movement in 1928 and had served as a revolutionary soldier through the grim days of the Long March and the epic guerrilla war against the Japanese. Liu was now a political commissar in a massive army that had just crushed the Nationalist forces assigned to defend Beijing. The Tsinghua delegation was headed by Zhou Peiyuan, scion of a wealthy landowning family in Jiangsu Province whose father had been a scholar-official under the Qing dynasty. Zhou had graduated from Tsinghua in 1924 and had studied physics at the University of Chicago and under Albert Einstein at Princeton, before returning to teach at his alma mater.[1] The delegations and their leaders were in many ways typical of the two very distinct groups who would face each other in the top echelons of Chinese society during the first decades of Communist power.

Red-over-Expert Power Structure

The CCP was founded by intellectuals, but during two decades of armed insurrection in the countryside it became a party of peasants. Its ranks were filled by poor villagers who took up arms in the anti-Japanese and civil wars, and even most of its leaders were of rural origin and had relatively little education. Robert North and Ithiel Pool, who analyzed changes in the party's top leadership during the decades it was fighting for power, documented how it was transformed by rural warfare. "Specifically, what was taking place was a rise in peasant leadership," they wrote. "The rise of Mao to power and the emergence of Soviet areas in the hinterland were accompanied by the replacement of intellectuals of middle-class and upper-class backgrounds by sons of peasants."[2] When the CCP took over China's cities, it was able to deploy a formidable corps of battle-tested cadres to take control of government offices, factories, and schools. These cadres were young, but many of them already had years of administrative experience in rural Communist base areas populated by millions of people. Almost all of them were from the countryside and few had much formal education. In 1949, 80 percent of the party's membership was of peasant origin and the great majority were illiterate or had only a grade-school education.[3] Those who had risen in the party's ranks were usually not from the poorest rural families, but rather from households that were moderately well-off by village standards and could afford to send at least one child to school; even the best educated among them, however, had rarely gone beyond middle school.[4] Although many of the top Communist leaders were born into elite families and had joined the party while studying at universities, even at the highest levels of the party a far greater number had come from more humble village origins and had risen to leadership positions by demonstrating their organizing abilities and military prowess.

During its first decade in power, the CCP eliminated the main foundation supporting the power and social standing of the old elite classes by systematically confiscating their productive property. This was accomplished through a series of mass political movements, which began in the countryside with Land Reform, a violent campaign in which Communist cadres mobilized poor peasants to humiliate, beat, and often kill landlords, and then redistribute their land. Landlords and rich peasants were not only dispossessed of their land and often their homes, but they were also reduced to social pariahs. The subsequent collectivization drive was less violent, but the result was more profound, eliminating private ownership of land altogether.[5] In urban areas, the state took over large enterprises, and small enterprises were combined into cooperatives. The process was largely peaceful, but fundamentally coercive. Communist cadres mobilized workers against their employers in a series of campaigns to combat tax evasion, corruption, waste, and counterrevolutionary activities, establishing Communist control within each enterprise and paving the way for state appropriation.[6] The fate of the urban elite, however, was not as dire as that of the rural elite because the CCP could not dispense with their expertise. The new government offered nominal compensation and management positions to entrepreneurs who cooperated, and the great majority of the managerial, professional, and technical staff in government offices, economic enterprises, schools, and other institutions remained in their posts. Members of the old urban elite nevertheless emerged from the early Communist campaigns greatly debilitated. They had been deprived of much of their property and were in a weak position politically. They retained other assets, however, that were highly valuable in a country that was largely illiterate—their education and expertise.

As the CCP took control of urban institutions, newly arrived Communist cadres were charged with supervising incumbent managers and specialists. In the parlance of the party, Reds were supervising experts. Bo Yibo, one of the CCP's senior leaders, described the encounter in his memoir. "It was natural that after we entered the cities, our core leadership in various fields was made up of cadres of worker and peasant origins who had just left the battlefields," Bo wrote. "These cadres mostly had a low educational level. They did not have much contact with the intelligentsia in the past, and did not know or understand the latter's professional expertise, mentality and working style."[7] The Communist victory had created a situation in which two very different groups coexisted uncomfortably at the top of the postrevolutionary social order: a new political elite, largely made up of peasant revolutionaries, and an old educated elite, largely composed of members of the dispossessed propertied classes.[8] Although there was overlap between the two groups, on the whole, they were of very different origins and had very distinct cultures and values. They also relied on different types of class resources—the former on political and the latter on cultural capital. Although practical considerations dictated cooperation, the first decades of Communist rule were marked by sharp conflict between the two groups.

ESTABLISHING COMMUNIST POWER AT TSINGHUA UNIVERSITY

In the education sector, the contrast between old and new elites was less pronounced because the party usually sent cadres with more education to take charge of schools. Nevertheless, at Tsinghua University the differences—in terms of social origin and level of education—between the newly arrived Communist cadres and the university faculty were quite obvious. Tsinghua's professors were a highly sophisticated and cultured group that included some of China's leading scholars. According to a survey conducted in 1946, over half the 134 professors and associate professors at the university had doctoral degrees, impressive in any country at that time and especially so in China. Almost all of them had studied abroad, mostly in the United States, and nearly half had degrees from Harvard, Massachusetts Institute of Technology, Cornell, University of Chicago, or Columbia.[9] The exceptionally strong foreign educational credentials of the Tsinghua faculty were a product of the university's history. Founded in 1911 by the United States using funds extracted from the Chinese government as part of the Boxer indemnity, Tsinghua was originally built as a preparatory school to train students to study at American universities.[10] The school was later taken over by the Chinese government and reorganized as a university, and during the 1930s and 1940s it ranked as one of China's leading institutions of higher education with a particular strength in science and technology. It continued to enjoy close relations with American universities, sending graduates to the United States for postgraduate training and then hiring them upon their return.

At that time, education was highly correlated with wealth, a fact amply demonstrated by the family origins of the Tsinghua faculty. When Communist officials conducted a survey of the class backgrounds of the university's professors, over 60 percent were classified as having originated from landlord or capitalist families, 27 percent were from professional or other middling categories, and less than 2 percent were from working-class or poor and lower-middle peasant families (a category that comprised over 80 percent of the population).[11] The group's impressive academic credentials and elite family origins were combined with political credentials that—in the post-1949 environment—did not help their standing. Not one was a member of the Communist Party, and many of them had ties to the defeated Nationalist Party. These connections, which had been assets under the old regime, were now severe political liabilities; those who had been members of the Nationalist Party—or even its youth organization—would now regret the association.

The team of Communist cadres that arrived to take over Tsinghua in 1952 was led by Jiang Nanxiang, a former Tsinghua student who shared the elite social origin of the university faculty (his family had owned a substantial amount of land in Jiangsu Province). Jiang had led student protests at the university against Japanese aggression in 1935, was expelled from the university for his protest activities, and then became a full-time cadre in the underground Communist movement. Before he returned to Tsinghua in 1952, at age thirty-nine, he had become a national deputy secretary of the Communist Youth League and he had many years of experience in the Communist underground. The party cadres Jiang brought to Tsinghua were in general younger, less well educated, and of humbler family origin. Many demobilized soldiers were assigned to administrative positions at the university. These worker-peasant cadres were typically of poor rural origin and the best educated among them had attended special accelerated middle schools established to train Communist cadres.

Jiang, who became university president and was later named secretary of the school party committee, was the dominant figure at Tsinghua for the next fourteen years. He eventually assembled a stable group of party committee leaders, all of whom had stellar revolutionary credentials. Two of Jiang's deputy party secretaries, He Dongchang and Ai Zhisheng, were former Tsinghua students who had been leaders of the underground party organization at the university.

The other four deputy secretaries were Communist veterans Jiang had brought in from outside. Of the veteran cadres, only Li Shouci, who had also been active in the anti-Japanese protests at Tsinghua in the mid-1930s, was highly educated. Moreover, at Tsinghua even Jiang and Li's educational credentials carried less authority because they did not have graduate degrees and they had been trained in the humanities instead of science and engineering.[12] The other three deputy secretaries, Liu Bing, Gao Yi, and Hu Jian, were of peasant origin and had only been to primary or middle school before joining the Communist movement (although they had received further training in party schools).[13]

On the whole, the Tsinghua faculty, who had run the university before 1949, did their best to accommodate themselves to the new regime. All were required to participate in political study meetings, in which they were urged to reform their thinking by breaking with "bourgeois ideology" and tendencies to "worship America." The Jiang administration appointed several sympathetic faculty members to leadership positions; a number of prominent professors, including Qian Weichang, Liu Xianzhou, Zhang Wei, and Chen Shihua, were made deputy university presidents and given positions on Tsinghua's administrative committee, formally the university's top governing body. Senior professors were also appointed as department directors, positions that traditionally had been vested with great power. As the CCP built branches in every university department, however, the real locus of power at all levels shifted to the school party organization.

The Institutional Foundations of Political Capital

With the conversion of the means of production into public property, access to advantageous class positions was no longer provided by economic capital (private property), but rather by political and cultural capital. Advantageous positions—whether in rural communes, state-owned and collective factories, schools, hospitals, or government offices—were now defined as cadre posts, and access to them required academic or political credentials. The former were more important for obtaining positions as technical cadres and the latter were more important for obtaining positions as political or administrative cadres. Academic credentials were distributed by the education system (discussed in Chapter 2), while political credentials were distributed by the party's recruitment apparatus.

THE PARTY ORGANIZATION

The value of political capital was underpinned by the extraordinary power of the Communist Party's organization, which commanded a bureaucratic apparatus extending from the top to the base of Chinese society. The party not only precluded political competition, but it also organized the entire populace around its political infrastructure. Rural villages were reorganized as collective production brigades and urban society was reorganized into Communist-style work units, all of which were led by a party committee or branch. Tsinghua was restructured in this fashion, and an examination of the university's structure will shed light on the nature of the party's power.

Communist leaders built a party organization at Tsinghua that paralleled the university's administrative hierarchy, and party committees and branches became the centers of decision making at all levels. Students, teachers, and other employees were all organized into small groups, and each group had a nucleus of party or Youth League members at the center. Teachers were assigned to "teaching and research groups" defined by academic specializations, students joined permanent classes, and university workers were organized into small teams. In addition to collectively organizing teaching, study, and work, these groups provided a site for political activities. Each group conducted its own affairs, but under party supervision, allowing for both active participation and effective social control. Both characteristics can be seen in the organization of student classes. A class of twenty-five to thirty students remained together during their years of study at the university; they took all of their courses together and lived in common dormitories. A teacher, typically a young party member, served as class director, and each class had a branch of the Youth League, which grew in size as new members were recruited. Youth League members elected a leadership committee, made up of a secretary and officers in charge of propaganda and organization, who organized political study and activities for all members of the class. In addition, the entire class elected a leadership committee, composed of a class president and officers in charge of study, recreation, labor, and cultural activities (such as theater and musical groups).

This type of small group organization was the key to the CCP's remarkable system of political control.[14] The party had a clear chain of command, in which individual members carried out the decisions of the party unit to which they belonged, and subordinate units implemented decisions passed down from above. The party was able to mobilize its members—and through them the entire population—because of the extraordinary commitment of those who joined the organization. Through this kind of thick infrastructural power, to use Michael Mann's term, the party was able not only to administer the university in a conventional sense, but also to mobilize the school population to carry out the Communist program of social transformation.[15]

THE POLITICAL CREDENTIALING SYSTEM

The CCP built a recruitment apparatus composed of an increasingly selective hierarchy of organizations that included, in ascending order, the Young Pioneers, the Communist Youth League, and the Communist Party. Young people were eligible to join the Young Pioneers at age nine, the Youth League at age fifteen, and the party at age eighteen, At each level, they faced more stringent requirements and more intense competition. By the early 1960s, almost all school children were invited to join the Young Pioneers, about 20 percent of the eligible age group were members of the Youth League, and about 5 percent of adults were party members.[16] These figures somewhat understate the proportion of young people who succeeded in joining the league and the party: some of those aged fifteen to twenty-five had not yet joined the league or had already left it to join the party, and young adults, who were the main target of party recruitment, were more likely to be party members than those who came of age before 1949. Nevertheless, both organizations made membership an accomplishment that required considerable effort.

Youth League recruitment was concentrated in senior middle schools, universities, and the military, all of which served as elite training centers. The young people selected to enter these exclusive institutions were very likely to join the league, where they would be initiated into the world of Communist activism. Young people who did not test into senior middle school or make it into the military, in contrast, could still apply to join the league in their villages or workplaces, but only a small minority did. Moreover, participation in league organizations in the mundane world of villages and workplaces was a much less intense experience than it was in the military and elite schools, which were both preparing future cadres. In rural areas, where few young people were able to attend middle school, military service (which was also very selective and conferred high status) became an important route for acquiring league and party membership. The heart of the recruiting effort, however, was in the school system.

The party's recruitment machinery became the organization's political nucleus at Tsinghua and other schools. While most of the university party organization was involved with supervising academic and administrative affairs, its recruitment apparatus was responsible for selecting and grooming young people to fill leadership positions, and it carried out much of the party's grassroots ideological and political education. A large corps of specially selected cadres was charged with carrying out these tasks, which were known as "student political thinking work," or "student work," for short. The student work apparatus took charge of the party's most political activities—publishing the university newspaper, managing the public broadcasting system, organizing political study campaigns, overseeing political courses, mobilizing volunteer labor, and recruiting new members.

The Youth League was at the center of student work, and senior league officers, who were party members employed by the university, were among the most powerful figures on campus. Among the cadres responsible for student work were politics instructors, who taught the required courses on party history and Marxist-Leninist philosophy, and teachers selected to serve as class directors, who worked closely with the student leaders of the league branches in their classes. President Jiang also initiated a system in which politically promising students were selected to become political counselors. Each counselor was responsible for supervising the political activities and recruitment of students in several classes.[17] Students selected to serve in responsible positions in the university league and party organizations, especially those appointed as political counselors, were being groomed for political leadership. Upon graduation, many were hired as political or administrative cadres at the university and elsewhere, and after the technocratic turn in the late 1970s some were promoted to top leadership posts. Among those who served as political counselors at Tsinghua during the Jiang Nanxiang era were Hu Jintao, who is now the CCP's secretary general, and Wu Bangguo, who is chairman of the National People's Congress.

Joining the Youth League was an arduous process that required proving oneself through years of political activity and volunteer work. Youth League branches in each class voted to accept or reject membership applications. The class director was responsible for writing summaries evaluating each student's political performance, and he or she consulted with leaders of the class league branch in preparing these summaries.[18] At elite schools like Tsinghua, a much larger proportion of the student body was recruited into the league than at nonelite schools; by the early 1960s most Tsinghua University students had already joined the league in middle school, and the great majority would join before they graduated from the university. In 1963, for instance, 84 percent of Tsinghua graduates had joined the league.[19] Party membership, in contrast, was a status that only a minority of students, who distinguished themselves through tireless political activism, would achieve. Among Tsinghua's 1963 graduating class, only 17 percent were party members.[20] Many others would apply after they had graduated and joined a work unit. Just over half of Tsinghua's junior faculty members, for instance, had joined the party by 1965.[21]

A complex mix of motivations inspired individuals to join the league and the party. As noted above, party membership was generally required for promotion to administrative positions, and by the early 1960s even technical cadres were expected to have at least achieved membership in the league.[22] Students, therefore, were clearly inspired by career calculations. It would be a mistake, however, to think that their motivations were simply instrumental. During this period, many students were deeply committed to Communist ideas and collectivist values, a commitment evinced in my interviews through nostalgic accounts, punctuated by contemporary slogans. Many students embraced Communist expectations and liked to think of their career ambitions as public service rather than personal advancement. For instance, Mei Xuesi, who joined the Youth League while studying at Tsinghua in the early 1960s, indicated that he and others were well aware that membership was important for their future careers: "If you wanted to achieve something," he explained, "you had to be in the Youth League and then the party." He insisted, however, that students' thinking then was not as instrumental as it is now. Students today, he told me, *only* join the league and the party in order to advance their career prospects. "Then, we didn't think about those things—we were so pure, we believed in the party. I didn't think of my career, I just wanted to be a good person, a good student, to get my job done well." Mei added that joining the league "was like a youth trend—you had to join to be in fashion."[23] Whatever the specific weight of career considerations, ideological beliefs, moral compulsions, and peer appraisal in an individual student's motivations, the result was compelling pressure to join the league. The few who did not succeed faced social isolation and a compromised future. This increased the stakes and enhanced the power of those who monitored the gates.

RECRUITMENT CRITERIA

In recruiting new members, the political criteria considered by party and league branches were divided into two categories. The first, political performance (*zhengzhi biaoxian*), was based on individual achievement, while the second, family background (*jiating beijing*), was ascriptive. Political performance was the most important criterion, and an individual's performance was evaluated in terms of three main elements: ideological commitment, collectivist ethics, and compliance with the authority of the party organization. Aspiring members were encouraged to diligently study Marxism-Leninism, and to demonstrate through their actions their commitment to the Communist ideals of hard work, selflessness, and a willingness to serve the people (*wei renmin fuwu*). These qualities were considered to be a reflection of an individual's moral quality.[24]

The main criteria for joining the Young Pioneers in primary school were cooperation, conformism, and compliance with authority. Tong Xiaoling, who attended the primary school attached to Tsinghua University in the early 1960s, described the required qualities as follows: "Don't fight, help other students, help the teacher, and be obedient."[25] Every year, red scarves were presented to new pioneers in an emotional induction ceremony. Age of admission was seen as a good predictor of a student's future political prospects, and only a handful of "problem" students failed to get red scarves before they graduated. Once students reached middle school, they were also required to gain fluency in Communist political doctrine. Ultimately, however, the Youth League was looking for the same moral qualities expected of Young Pioneers. "To join the Communist Youth League," Mei Xuesi explained, "you had to be a good student, work hard, be a good helper, [do volunteer] work in the countryside, and don't say anything bad about the teacher, the country, or the party."[26]

In party history courses, students learned about the heroic exploits of Communist revolutionaries and they were encouraged to follow in their footsteps by joining the league and the party. In the years since 1949, however, political activism had lost its subversive quality and became a more domesticated endeavor. The characteristics most prized in activists were now loyalty to the new government and compliance with authority. Liu Shaoqi, the party's organizational chief, made this clear in a series of lectures on the personal conduct expected of Communist cadres that were published in a thin volume titled

How to Be a Good Communist, which was required reading for those wishing to join the party.[27] Compliance with authority did not mean abject servility by any means. On the contrary, the party was seeking individuals who displayed an ability to work effectively in a bureaucratic hierarchy, both accepting guidance from above and providing direction to those below.

Teachers at Tsinghua and its attached middle and primary schools did not find the qualities elaborated in Communist guidelines for teaching morality to be completely unfamiliar, as many of the basic themes were not new. In Confucian tradition, proper moral training had long been considered an essential part of cultivating honest, benevolent, and loyal state officials, and the considerable benefits enjoyed by imperial officials came with a moral obligation to dedicate themselves to public service. During the late imperial and Republican periods, modern patriotism was grafted onto these older Confucian ideals. After 1949, moral education classes continued to promote all of these themes, although the Communist regime gave them a new ideological content.

The other major consideration in recruitment—family background—involved two sets of categories: class origin (*jieji chushen*) and political background (*zhengzhi beijing*). All families were assigned a class origin designation according to the CCP's taxonomy of classes (see Figure 1.1). These designations were based on the status of the family head between 1946 and 1949, and were inherited patrilineally during the first three decades after the 1949

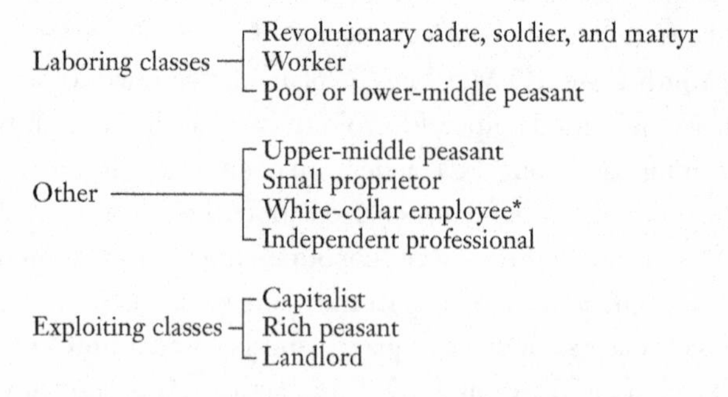

Figure 1.1 Class origin categories

*The category *zhiyuan*, which I have translated as "white-collar employee," included government officials, office workers, teachers, and managerial, professional, and technical employees. These occupations required a relatively high level of education, and the number of *zhiyuan* was very small relative to the population of prerevolutionary China. Their class position was, therefore, higher than that of clerical workers in developed countries.

Revolution.[28] Along with class origin, the political history of an individual was evaluated, and those determined to have committed counterrevolutionary or criminal offenses faced discrimination, as did members of their families.[29]

Class line (*jieji luxian*) policies used these designations to give preference in political recruitment (as well as in school admissions and job appointments) to members of working-class, poor and lower-middle peasant, and revolutionary cadre, soldier, and martyr families. The overwhelming majority of the Chinese population belonged to these categories and most of the rest had middling designations that did not provide advantages, but also did not present great handicaps. For members of the small minority of families deemed to be part of the former exploiting classes or associated with counterrevolution, however, class line discrimination could have devastating consequences. This was especially true in the countryside, but even in the cities, where former capitalists and officials of the old regime often retained respectable positions, individuals with problematic class and political labels faced discrimination and were particularly vulnerable during periods when class line policies came to the fore.[30]

Class line policies turned the class hierarchy that existed before the 1949 Revolution on its head, providing advantages for those who had been at the bottom and disadvantages for those who had been at the top. They were based on two related political and social rationales. First, the party wanted to promote to leadership positions individuals from social groups considered to be protagonists and beneficiaries of the Communist project, and to prevent these positions from falling into the hands of individuals associated with social groups considered to be hostile or potentially hostile to this project. Second, the party wanted to counter the substantial cultural advantages enjoyed by members of more educated families by providing political advantages to those whose educational opportunities had been more limited.[31]

These policies greatly benefited members of the new political elite and hurt members of the old educated elite. Family origin, however, was only one consideration in recruitment. The party was determined to recruit capable and committed new members, and party and league branches were instructed to look first at political performance. Children of revolutionary cadres had great advantages in the competition to join the league and the party, and party officials were often very solicitous toward these students, encouraging early membership applications and providing opportunities to assume leadership responsibilities. Nevertheless, even these students were expected to earn their political credentials through hard work, and indications of aristocratic indolence were criticized. Students from old elite families that had been stuck with bad class labels were also invited to join the league and the party, but they faced particularly stringent tests of their political reliability. It was possible to overcome these handicaps, however, and many excelled in the political competition.

Bureaucratic Authority, Personal Dependency, and Social Hierarchy

During the Jiang Nanxiang era, the Tsinghua party organization was referred to as an "engine that leaks no steam" (*yong bu louqi de fadongji*). Jiang created a remarkably monolithic and efficient organization by removing opponents—including both nonparty faculty members and party officials—from positions of power, and replacing them with individuals who were loyal to him. At Tsinghua, as in all Chinese work units, decisions about hiring and promotions were handled by the party organization. Chinese universities generally preferred to hire their own graduates, and Jiang was particularly insistent on hiring Tsinghua graduates. He was convinced that the university provided the best technical training, but he also preferred to select teachers and cadres from among those groomed in his own political stables. Tsinghua graduates who were tapped to become administrative and political cadres at the university were referred to as "Tsinghua-brand cadres" (*Qinghua pai ganbu*), and they were known for their strong loyalty to Jiang and the university party organization.

During the early 1960s, when Jiang served concurrently as the president and party secretary at Tsinghua and as China's minister of higher education, he enjoyed unchallenged personal authority at the university. He was a demanding leader with a strong personal will and a penchant for order and discipline. His motto, addressed to the entire university community, was: "Be obedient and productive" (*tinghua chuhuo*). "You couldn't disagree with him," said Zhuang Dingqian, a mid-level Tsinghua cadre who retains a profound respect for the late university president. "Jiang Nanxiang's authority was very strong at Tsinghua—everyone listened to him."[32] Although Jiang's tight control stifled debate, the orderly and disciplined atmosphere also fostered a certain kind of dynamism at the university, and even those who had chafed under Jiang's political tutelage admitted that in the early 1960s morale at the university was high. Teachers and students worked extremely hard, inspired by intense political and academic competition and a widely shared ethic of public service. Mai Qingwen, a university party official, recalled the period with nostalgia. "Under Jiang's leadership, control at Tsinghua was very strict," he said. "Cadres were not corrupt; there were no power struggles; they didn't attack each other; everybody worked for Tsinghua's future."[33]

Jiang's personal authority and the power of the university party organization were completely intertwined, and the party's formal hierarchy became a trellis on which informal personal networks flourished. Although these personal ties in some ways hindered the impersonal bureaucratic rules of the organization, they also reinforced authority relations between superiors and subordinates, and they became an integral part of an emerging social hierarchy. The power concentrated in the hands of party officials facilitated the development of relations of dependency with their subordinates, which in turn reinforced officials' power. The party's practice of rewarding loyalty to the organization was inevitably used by Communist cadres to reward personal loyalty among their subordinates.[34] Such clientelist patterns were established in the earliest stages of the recruitment process, because this process made loyalty a key selection criterion and it fostered a style of activism characterized by conformism and compliance with authority. This kind of compliant activism became an essential feature of the party's political culture.

From a state-building perspective, the party organization was a highly effective instrument of social control, and it gave the new regime a tremendous capacity to mobilize the population to carry out its program. The political credentialing system played a critical role, allowing the party to select and groom future leaders and to infuse young people with Communist values and ideology. At the same time, the system became the object of individual strategies to get ahead. Because the credentials distributed by this system—membership and leadership positions in the league and the party—were required to win jobs and promotions in the state bureaucracy, they became a very tangible, institutionalized form of political capital. More informal and less tangible, but just as important, were the personal networks derived from association with the ruling party.

By extending and consolidating the party's power, Communist officials shored up their own elite status. Their positions in the party hierarchy provided them with personal political capital and gave them opportunities to pass their advantages on to their children. The party, however, was not a closed organization, and children of the old elite classes as well those of more humble origins were diligently participating in the party-sponsored political credentialing competition. Thus, the party organization and its recruitment apparatus became the political foundation of an emergent class order. In fact, with the elimination of private property in the means of production, it became *the* most important mechanism of class differentiation in postrevolutionary China.

Challenging the Political Hierarchy

In the spring of 1957, Mao Zedong invited nonparty members to join in a campaign to Rectify the Party's Style of Work. The campaign, initiated just as the party was celebrating the completion of China's transition to socialism (with the elimination of private property in agriculture, commerce, and industry), was Mao's first major effort to address a problem that would preoccupy him for the remainder of his life—the emergence of a privileged elite made up of Communist officials. In launching the campaign, Mao highlighted three problems—bureaucracy (*guanliao zhuyi*), sectarianism (*zongpai zhuyi*), and subjectivism (*zhuguan zhuyi*). In the Chinese Communist lexicon, bureaucracy referred to cadres concentrating power in their own hands, sectarianism referred to cadres alienating themselves from those who were not in the party, and subjectivism referred to cadres making decisions based on narrow knowledge and considerations. All three problems were perennial concerns that Mao had made the targets of previous party rectification campaigns. This time, however, the prescribed methods were different. In the past, people had been invited to criticize local Communist cadres, but always in forums orchestrated by the party.[35] Now people were invited to air their complaints in a more spontaneous manner, and this kind of unrehearsed, unsupervised criticism was unprecedented.

The 1957 Party Rectification campaign was a continuation of the campaign to Let One Hundred Flowers Bloom and One Hundred Schools of Thought Contend (*baihua qifang baijia zhengming*) launched a year earlier, but now Mao requested that criticism be directed specifically at party practices and party officials. To the chagrin of other party leaders, Mao extended the invitation especially to intellectuals, who staffed the offices of enterprises, schools, and government departments, working under the supervision of veteran Communist cadres.[36] Mao's appeal to intellectuals to help rectify the party provoked a storm of conflict between China's political and educated elites that had been building during the first eight years of Communist power. First, intellectuals seized the opportunity to criticize party officials, and then party officials struck back with devastating effect. The political credentialing system became a focal point of the debates, with critics challenging the value of political qualifications and denouncing the conformism fostered by the system.

In the years leading up to the 1957 campaign, Mao had often expressed his dissatisfaction with the stiff and orderly style of politics practiced by the party since it had taken power, and he had a particular distaste for scholastic means of teaching the principles of communism. "If the lectures in politics classes are so boring that they make people go to sleep," he suggested, "then it would be better not to lecture and instead let people sleep and save up their energy and spirit."[37] Nostalgic for the years of revolutionary upheaval, Mao preferred that young people learn politics directly through political struggle. The orderliness and conformity that Tsinghua's president and other Communist leaders prized made Mao uneasy. He favored open contention and this is what he demanded in the spring of 1957. "The meaning of the Central Committee," Mao told party officials at the outset of the Party Rectification campaign, "is that you can't control everything, you have to let go (*bu neng shou, zhi neng fang*); just let go, and let everyone express their opinions, let people speak out, criticize, and debate."[38] When Jiang Nanxiang resisted this demand and continued to prevent students at Tsinghua from holding meetings, Mao reportedly told him: "There's nothing to be afraid of. If you can't hold Tsinghua University, you can withdraw to Dongchang'an Street; the People's Liberation Army is there. If you can't hold Dongchang'an Street, then you can retreat to Shijiazhuang, or to Yan'an."[39]

During May and early June of 1957, the Party Rectification campaign came to life at Tsinghua. Freed from the tight political control that had characterized Communist governance at the university since 1952, students and teachers organized free speech forums (*ziyou luntan*) in departments and large outdoor meetings to criticize university officials and policies. They plastered school walls with provocative big character posters (*dazibao*), handwritten manifestos that had become an important form of public political discourse. Students gathered names on a petition demanding that Tsinghua be reorganized as a comprehensive university, reversing the Communist restructuring that had made Tsinghua into a Soviet-style engineering school (see Chapter 2). Students and teachers from nearby universities visited Tsinghua, and

Tsinghua students and teachers went to other campuses to witness the polemics unleashed after Mao had called for "free speech and great debates" (*daming dafang dabianlun*).

CHALLENGING THE VALUE OF POLITICAL QUALIFICATIONS

During the Nationalist era (1927–49), the Tsinghua faculty had resisted government interference in the university and Mei Yiqi, who served as university president during most of that era, had championed university autonomy under the motto, "Schools should be run by professors." Now the CCP not only insisted on running the university, but also on radically transforming it according to the Soviet model. Although President Jiang had appointed senior professors to positions of authority, they found real power was steadily shifting to the school party organization. The new power dynamics inverted the traditional status order in ways that were jarring for faculty members, especially senior professors. First, they had to listen to Communist cadres who were much younger, and second, the education level of many of these cadres was far inferior to their own. Although most Tsinghua faculty had made great efforts to accommodate themselves to the new regime, many bristled at the changes it demanded, and most had difficulty reconciling themselves to the Red-over-expert premise of the new order. In 1957 they questioned this premise, making the relative value of political and academic qualifications a central issue in the ensuing debates.

The two leading antagonists were Jiang Nanxiang and Qian Weichang. Both Jiang and Qian studied at Tsinghua in the 1930s and both participated in the historic "December 9" student movement against Japanese aggression in 1935. Afterward, Jiang left to join the Communist underground; when he returned to the university in 1952, he had long since abandoned the traditional silk scholar's gown he had worn as a student and had grown accustomed to the plain cotton uniform of the revolutionary cadre. Qian instead went to Canada and then to the United States, where he studied rocket design at the California Institute of Technology, and in 1946 he returned, wearing a Western suit, to take a teaching position at Tsinghua. In 1949, Qian was among the professors who staunchly supported the new regime, and he encouraged young scientists and engineers to return to Tsinghua from abroad to help build a strong university and a strong country. Qian was appointed to serve as deputy president of the university and dean of academic affairs, and, as a renowned scientist who was sympathetic with the government, he was invited to play a prominent role in discussions of national science policy, rubbing shoulders with top Communist officials.

Jiang and Qian voiced the competing claims of Communist officials and veteran professors to exercise power at Tsinghua. Yue Changlin, a senior Tsinghua professor who became one of the sharpest critics of the party in 1957, succinctly described the conflict between Jiang and Qian. "After 1949, no professor was a real leader of the university; even if they were leaders in name, they had no actual power," Yue told me. "That's why Qian Weichang struggled with Jiang Nanxiang. . . . Qian was the dean of academic affairs, but he wanted to lead the school. . . . Jiang was just a bureaucrat; Qian insisted that the leader of the school should be a scientist. Jiang demanded Communist Party control; he insisted that politics was the key issue. . . . They were actually both fighting for power."[40]

Qian had fully cooperated with the Communists, but he resented party control, and he now criticized party officials for pushing scholars out of the decision-making process. "In recent years," he complained, "we have less and less of a feeling of being masters."[41] A man of great self-confidence known for speaking his mind, Qian brashly articulated what became a key refrain of intellectuals—that "nonexperts cannot lead experts" (*waihang bu neng lingdao neihang*). "We need people who have professional knowledge to express their opinions; we shouldn't let those who have no professional knowledge spout nonsense (*wawa jiao*)," Qian declared. "Debates should be carried out among those who are academically qualified; promotion should also be based on academic standards. It is wrong to promote people based on mass support and how well they can talk the talk. . . . By academic leadership I don't mean that old men must lead. It doesn't matter if you are old or young, as long as you have academic qualifications, you can lead."[42]

Qian and other veteran faculty members criticized the party for bringing to the university rural origin cadres with relatively little formal education and placing them in positions of authority over highly educated professors. Liu Bing and Hu Jian, the deputy secretaries responsible for the party's organizational affairs, were singled out for criticism. Liu and Hu were both of peasant origin; Liu had not finished middle school and Hu had only been to primary school (although both had participated in Communist training classes in Yan'an). They were, in the words of angry faculty members, "country hicks" (*tubaozi*) who had no business leading a university.[43] Veteran faculty members were also upset that young party members, many of whom had only recently graduated, were being placed in positions of authority at the university and department level, and were now giving orders to their former professors. Senior professors, even those who were nominally department directors, were being left out of decision making, which increasingly took place in department party branches. Moreover, they complained, the selection of students and young teachers for special training and promotion was based more on political than academic criteria.

Tsinghua professors also resented being required to teach special classes made up of party cadres of worker and peasant origin who had little previous education. "Workers' and peasants' classes are also another typical sectarian method," declared Professor Tang Shuxiang at a forum organized by teachers in the electronics department. "They bring worker and peasant party members here to study, and we spend an awful lot of effort to train them, but the results are still bad."[44] A salient theme articulated by Qian and other party critics was that young party and league members—both students and young teachers—did not show the proper deference to professors their senior in age, professional rank, and knowledge. Qian complained that the Communists were ruining teacher/student relations. "You want to help them study, but they have a skeptical and critical attitude in learning from you; even worse they want to struggle against you," he wrote. "How can anyone teach under these conditions? How can anyone learn?"[45]

When Qian and other professors argued that it was wrong for nonexperts to lead experts, they were directly challenging Communist pretensions to authority at the university. Both sides, however, recognized that the implications were much broader. The debate was over the relative importance of academic and political qualifications in selecting people to fill decision-making positions in all sectors of society, and members of the educated classes were voicing a deeply held belief that they were better qualified to lead society than the Communist usurpers.

DENOUNCING POLITICAL PRIVILEGE AND CONFORMISM

In addition to criticizing political intrusions on the prerogatives of cultural capital, the party's intellectual critics also condemned the Communists' tight political control and warned that the new system was giving rise to social differentiation based on political affiliation. The language they used was shaped by the narrow ideological boundaries the party imposed on the discussion, which were respected by most of those who spoke out. They were aware that their criticisms would be most effective when presented within the prevailing political paradigm and, in fact, Communist rhetoric about democracy and social equality provided ready templates for poignant critiques of the party's practice. A rank-and-file member of the Youth League pointed out that elections for committee positions were often like arranged marriages because the leadership imposed their favorites, and he criticized the political counselor system as bureaucratic because it concentrated power in the hands of a few individuals who were appointed from above.

Faculty members criticized an emerging status hierarchy connected with political affiliation. One noted that party members and those who were applying to join the party all sat together at the same tables in the cafeteria, refusing to mix with the masses. Another complained that his students no longer even greeted him after they joined the party. Becoming a party member, he said, seemed tantamount to moving a rung above other people and joining a distinct class. In a similar vein, Professor Ding Zeyu criticized the system of preferential treatment cards, established in 1956, which provided senior professors and cadres with special access to health care and other goods and services, slyly pointing out that rank was replacing wealth in providing privilege. "In the past, the Nationalists had advantages over the masses because they had money, but they never issued preferential treatment cards."[46]

Qian Weichang forcefully criticized the party's culture of conformism and its stultifying effect on Communist cadres. "Today the cadres they train are all obedient—in fact, they only act when you push them," he declared. "They have no ability to think independently. Only those people who do not use their brains can get along with the party."[47] Further developing this theme in a speech published in the national daily, *Chinese Youth*, Qian condemned Communist efforts to impose ideological and organizational fetters on young people. "When they talk about educating youth, they mean an unlimited number of rules, drawing lines to confine youth," he wrote. "Why is education not seen as training and reason, but rather as limitation and control? This has to do with the low level of the cadres, but more important it reflects the remnants of feudal thinking. There was a whole set of feudal sayings about how to 'educate' youth, requiring 'maturity and prudence' and 'refinement and cultivation.' Today, although we don't use this language anymore, some of the content has been reincarnated in sayings like 'submit to' and 'humbly listen to the opinions of the masses.'"[48]

By early June, critics of the party were becoming increasingly bold. At free speech forums held at Tsinghua, party leaders were denounced as "local tyrants," "fascists," and members of a "privileged class," and party and league activists were referred to as their "running dogs." One professor compared the Communists to the Manchurian rulers of the Qing dynasty, a metaphor that recalled nationalist narratives about Han civilization being overrun by barbarians. Speakers called for the abolition of the party committee system and even demanded that the party withdraw from the school. Others suggested that the party "return state power to the people."[49]

ANTI-RIGHTIST COUNTERATTACK

In mid-June 1957, after enduring six weeks of increasingly sharp criticism, party officials retaliated with a vengeance, launching a campaign against "bourgeois Rightists." Deng Xiaoping, the CCP's general secretary, organized the campaign and Mao endorsed it. Vocal critics were denounced as champions of the old exploiting classes who wanted to reclaim power and retain their monopoly over knowledge and the privileges this entailed. Using top-down methods of mobilization perfected in previous political movements, party and league activists were called upon to defend the party. At Tsinghua, department and schoolwide meetings were held to denounce teachers and students accused of being Rightists and school walls were covered with a new layer of big character posters. This time, however, the content of the speeches and the posters was carefully orchestrated by the party organization. In a front-page article in the school newspaper, several students of Meng Zhaoying, the director of the electronics department who had been denounced as a "Big Rightist," criticized him for disparaging party members as "yes-men with wooden brains" (*weiwei nuonuo de munaogua*). "A party member must obey the party's decisions and the instructions from above, that's the only way to maintain the party's fighting power," they argued. "If party members all followed Meng's advice and questioned the party's demands, what would happen to the unified action of the party? Wouldn't that destroy the party's thinking and organization?"[50]

Meng and other vocal critics of the party were made to pay a heavy price for their temerity. Across the country, hundreds of thousands of people were punished as Rightists, including 403 students and 168 teachers and other employees at Tsinghua.[51] Many teachers were removed from their posts and some were sent to labor in the countryside. Qian Weichang, Meng Zhaoying, and several other senior professors were removed from the administrative positions they held. Those promoted to replace them included senior professors who were more reticent during the Party Rectification movement and proved their loyalty during the subsequent Anti-Rightist campaign, as well as young party members. He Dongchang, who had been an underground student leader at the university before 1949, was tapped to take over Qian's duties overseeing academic affairs.

The Anti-Rightist campaign also targeted many party and league members who had spoken out during the Party Rectification. A subsequent campaign in 1959 specifically targeted "Right deviationism" in the party; eleven party members were denounced as Right deviationists, fifty-three were said to have displayed "serious Right deviationist thinking problems," and 146 members were singled out for "help."[52] Eventually, almost all the teachers who had been punished, including Qian Weichang, were allowed to return to teaching, and by 1964 all but ten of the 168 Tsinghua employees who had been denounced as Rightists were formally rehabilitated. Nevertheless, everyone who had been stigmatized, including those whose Rightist "hats" were formally removed, continued to face discrimination until after the end of the Cultural Revolution decade.

The Anti-Rightist campaign made it clear that criticism of party officials and party policies would not be tolerated. Speaking before a school meeting in 1958, Jiang Nanxiang warned, "[The Rightists] underestimated other people, the masses, and the party; they lacked the attitude of seeking truth from facts and were not sincere and well behaved. They thought they were really something and could play with others. . . . The Rightists should not overestimate themselves in the future—they'll lose in the end. This experience should be studied in order to develop a well-behaved attitude (*laolao shishi de taidu*)."[53] The impact on members of the university community was profound. Students and teachers who witnessed the punishment of those who had spoken out learned to measure their words carefully. Yang Yutian, who had just begun studying at the university in 1957, recalled, "The impact on us was—don't talk carelessly or recklessly; you have to do what the party says (*ting dang de hua*)."[54] Students who arrived at the university in the wake of the Anti-Rightist campaign were eager to keep their distance from the controversy. Wei Xuecheng, a young woman who arrived at Tsinghua in 1958, told me that students in her cohort were not interested in criticizing the party. "We thought people should listen to the party; we wanted to be obedient tools of the party."[55]

Why did Mao in 1957 first call on intellectuals to criticize the party, and then turn on those who heeded his call? One common answer is that he and other Communist leaders were simply playing a trick on intellectuals by encouraging those with dissenting opinions to speak out so they could be identified and suppressed.[56] Others have suggested more complex reasons. Roderick MacFarquhar argued that Mao's decision to invite unsupervised criticism of the party was strongly opposed by most party leaders and he was compelled to beat a hasty retreat after criticism got out of hand.[57] Merle Goldman proposed that this episode was one of a series of cycles in which the party first relaxed control over intellectuals in order to encourage their creative participation in solving problems it faced, and then repressed dissidence in order to re-impose political control.[58] Because this book does not delve into biography or the details of elite party politics, I cannot arbitrate between these interpretations. If we consider the events of 1957 in light of the persistent conflicts over political and cultural power during the Mao Zedong era, however, three general observations can be made. First, the bureaucratic power of party officials—which Mao made a target of the 1957 Party Rectification campaign—was a problem he returned to again and again, each time with greater virulence. By soliciting unsupervised criticism by intellectuals, Mao opened the way for a much more profound discussion of this problem than would have been possible had the party been asked to rectify itself. Second, Mao was completely unsympathetic with intellectuals' conviction that their expertise made them more fit to run the country than the Communists, and much of the most hostile invective directed against intellectuals during the Anti-Rightist campaign reflected his thinking. Third, Mao was a master of Machiavellian manipulation of contentious political forces for his own ends, and he was probably not dismayed by the sharp conflicts between new and old elites produced by the Party Rectification campaign. These points, however, can only be fully discussed once the dramatic events that unfolded a decade later, during the Cultural Revolution, are brought under scrutiny. In this subsequent upheaval, the antibureaucratic themes of 1957 would be brandished by a movement with a much wider class base and much greater destructive power.

In any case, the events of 1957 displayed in sharp relief the antagonisms between the country's new and old elites. The Party Rectification campaign gave members of the old educated elite a chance to vent their grievances about experts being supervised by Reds, and they enthusiastically challenged the authority of the party organization and denounced the power and privileges of the new political elite. In the subsequent Anti-Rightist movement, the party organization responded with a vindictive counterattack that reasserted its authority and discouraged further dissent, entrenching the political foundations of a newly emerging class order. This movement was also the beginning of a sustained assault on the old elites and the cultural foundations of their power, which will be the topic of the next chapter.

Chapter Two

Cultural Foundations of Class Power

On October 25, 1952, Tsinghua University convened its annual assembly to welcome new students, but this year the celebration was extraordinarily grand, with top party and government leaders in attendance. The event marked the opening of a reorganized higher education system, transformed in line with Soviet principles. Tsinghua had been converted from a comprehensive university into a polytechnical industrial university, moving it to the forefront of the Communist industrialization drive. Moreover, the entering class of 1952 was the first cohort of students to be selected by means of a unified national entrance examination. Tsinghua, like other universities, had traditionally given its own entrance examinations; now recruitment was centralized and standardized. As a national-level university, Tsinghua was given a quota of students to enroll from all of China's provinces, ensuring that its graduates would represent the entire country. The one thousand or so new students who crowded into the university's largest auditorium could be proud that they had been selected from among applicants with the very highest examination scores in the country.

Almost all of these students were from old elite families. In 1952, with the inauguration of the national examination system, the government also carefully recorded for the first time the class origin of university students. From a class line perspective, the results were disappointing: of the 3,160 students enrolled at Tsinghua, only 14 percent were from working-class or peasant families.[1] The poor showing of children from humble families was not surprising; in the early 1950s, less than 1 percent of young people graduated from senior middle school, and the great majority of those who did had been born into educated, well-to-do families.[2] Moreover, during the fourteen years that this national examination system was in place (1952 to 1965), it continued to predominantly select children from a tiny minority of educated families, methodically reproducing the cultural advantages of the educated elite.[3] Despite Communist efforts to rapidly expand primary and secondary education, as long as admission was regulated by examinations, most students who tested into Tsinghua and other top universities were from the old elite classes.

Between the 1949 Revolution and the Cultural Revolution in 1966, education policy oscillated between technocratic and radical egalitarian orientations. On the one hand, the new regime created a highly centralized, hierarchical, and meritocratic education system modeled closely after the contemporary Soviet system, which by then was organized according to technocratic principles. The system was designed to quickly train a large corps of technical experts to aid in the Communist industrialization drive. On the other hand, the CCP was resolutely committed to eliminating class differences based on education. Chinese Communist leaders had also inherited this commitment from their Soviet mentors. In Marxist-Leninist theory, after the conquest of political power and the socialization of the means of production, Communists still faced the task of eliminating the "three great differences"—between worker and peasant, city and countryside, and mental and manual labor. The division between mental and manual labor, which in Marx's grand historical narrative arose as early human societies first divided into classes, would have to be overcome before a socialist society could advance to communism. During the early years of the Soviet regime, the Bolsheviks had made the elimination of the distinction between mental and manual labor a fundamental aim of the socialist state, and under this banner they had pursued extremely radical education policies.[4]

By the 1950s, the Soviet Union had long abandoned these policies in favor of technocratic policies, but Mao Zedong and other Chinese leaders became increasingly disturbed by the elitist nature of the Soviet model. They not only remained committed to the Marxist doctrine of class leveling, but they also deeply distrusted the old educated elite and were determined to prevent them from reproducing their class advantages. This distrust was shared by a good part of the party leadership and membership, most of whom had little education. By the late 1950s, Mao led the CCP to reject much of the Soviet model and instead pursue a radical program to "revolutionize education" and undermine the social position of the educated classes.

Restructuring the Academic Credentialing System

During its first decade in power, the CCP reorganized the school system from top to bottom. Private schools were taken over by the state, all schools were folded into a national administrative hierarchy, and the nationally coordinated examination system was extended to regulate school admission at all levels. In the fourth grade of primary school, students took exams that determined who among them would continue on to the fifth grade (senior primary school), and those who succeeded faced further examination hurdles to enter junior middle school, senior middle school, and college. The school system was shaped like a pyramid, with very few places at the upper levels, and competition was fierce.[5]

The new regime greatly increased access to education by rapidly expanding the school system, especially at the basic level. In 1949, the majority of children did not attend school at all, less than 7 percent completed primary school, only about 2 percent completed junior middle school, less than 1 percent completed senior middle school, and an even smaller proportion went to college. Seventeen years later, on the eve of the Cultural Revolution, almost all children finished junior primary school, about 36 percent finished senior primary school, over 10 percent finished junior middle school, about 3 percent finished senior middle school, and over 1 percent attended college.[6] Even after this impressive expansion, however, the education system continued to distribute knowledge and academic credentials very unevenly and most children—especially in the countryside—were excluded after a few years of schooling.

The Communist regime also significantly changed the content and the categories of knowledge imparted by the school system. In 1952, higher education was reorganized in line with the Soviet model and most colleges and universities were assigned specialized teaching missions with a narrow range of academic disciplines.[7] The reform had a tremendous impact on Tsinghua, which had been modeled after comprehensive universities in the United States and now became a Soviet-style polytechnical industrial university. Tsinghua's humanities and science departments were transferred to Peking University and its agriculture and aircraft design departments were transferred to nearby specialized colleges, while its architecture and engineering departments were reinforced with teachers transferred from other schools. The structure of academic disciplines was also changed to match the Soviet system, as were teaching methods, assessment procedures, curricula, and materials. Before 1952, Tsinghua had proudly followed American practices, even borrowing elements of campus culture, including sports competitions and hazing. The curricula and textbooks of many engineering courses were borrowed from the Massachusetts Institute of Technology and other U.S. universities, where faculty members had received their training. Virtually all teachers and students read—and most spoke—English, and professors accented their lectures with English words and phrases.[8] Now, course designs were imported from the Soviet Union and English-language texts were discarded in favor of Russian texts or hurried translations.

While the American model had provided a broad general education, the Soviet model was geared to produce highly specialized engineers. It corresponded with the Soviet system of unified job placement, also adopted by China, in which college graduates were directly assigned to positions connected with their specializations. The Soviet system of specialized training implemented at Tsinghua was narrow, rigid, and highly rigorous. Students were assigned to a particular major from the time they arrived at the university, they all took a prescribed series of the courses, and the content of the courses was defined in meticulous detail. Along with Soviet pedagogy, the CCP imported a hierarchy of rationally defined academic and professional ranks, access to which was governed by academic credentials. The meritocratic principles that underpinned this system were tempered by collectivist principles, as Soviet education philosophy stressed the needs of the group and the organization. It did not encourage stars and it resisted the division of students into different classes based on ability. "The Soviet system provided for all students to reach a standard level by graduation," explained Tong Yukun, a top Tsinghua leader. "Perhaps it didn't allow as much creativity and it didn't allow the best to excel to their capacity; but it limited the gap between the top and the lesser students."[9]

Soviet-inspired education reforms were designed to support a program of rapid industrialization also modeled on the Soviet experience and undertaken with Soviet help. Marshaling and allocating resources through centralized planning, the new Chinese regime created whole new industries, transformed small cities into massive industrial centers, and built railroad and electrical grids that connected most of the country. Industrial production grew by an average of 11.5 percent a year between 1952, when the First Five Year Plan was launched, and 1978. During the same period, industry's contribution to the gross national product grew from 18 percent to 44 percent.[10] Rapid industrialization required large numbers of engineers, and the CCP greatly strengthened engineering training in Chinese universities, which had previously been weak. Between 1947 and 1965, the number of Chinese university students grew nearly fivefold (from 130,715 to 644,885), and, even more remarkably, the number of university students studying engineering grew more than twelvefold (from 23,035 to 292,680).[11] Moreover, even though there were few engineering graduates before the Communist era, not all of them could find engineering jobs. Now, despite the large increase in the number of engineering graduates, their services were in great demand. All graduates from universities, colleges, and technical senior middle schools were guaranteed placement in cadre jobs.[12] Job assignments corresponded with the rank of the school, and because Tsinghua was a top-ranked national university, its graduates typically received prized assignments in national and provincial government offices, industrial ministries, large industrial enterprises, research institutes, and universities.

Thus, during the first decade of Communist power, the academic credentialing system was centralized and rationalized, given a more technical orientation, and greatly expanded, so that it was now more open to children from working-class and peasant families. Moreover, the Soviet imprint gave the credentials it offered a socialist legitimacy. Nevertheless, it remained a powerful instrument of class differentiation that continued to facilitate the reproduction of the old educated elite. The education reforms of the 1950s presented significant challenges to Chinese intellectuals who had been trained before 1949. Many of them were fluent in the literary traditions of classical Chinese, English, and other Western languages, which they had acquired through years of study in Chinese and Western universities. After 1949, mastery of these cultural traditions continued to command great respect, but they also attracted suspicion as markers of the "feudal" and "bourgeois" culture, and the value of English and classical Chinese fell sharply, as the texts that contained the keys to intellectual advancement—whether in terms of technique or philosophy—were now inscribed in modern vernacular Chinese or in Russian. While struggling with new linguistic and cultural demands, intellectuals also had little choice but to abandon the traditional long gowns and the Western suits that previously had distinguished them from the less educated classes. All of these changes significantly altered the conventional categories and customs that defined cultural capital.

Many members of the Tsinghua faculty were deeply disappointed with the conversion of their university into a polytechnical school, which in their eyes had reduced its status, and they were uncomfortable with the Soviet curricula, materials, and teaching methods. Nevertheless, they adapted to the new conditions. Tsinghua professors learned the ascendant languages of Russian and Marxism and they adjusted to the narrow contours of Soviet engineering specialties, making them their own. Students at Tsinghua and other universities, most of whom had come from old elite families, were in a better position than their parents to embrace the new expectations and opportunities. Many thrived in the overheated academic environment oriented to the Communist industrialization drive. They used the cultural resources at their disposal to master the goals presented by the new examination system and the rationalized hierarchy of professional ranks.

Although members of the old elite classes were forced to surrender their propertied wealth and many now worked under the supervision of revolutionary cadres, they nevertheless retained their cultural advantages, and the Communist regime continued to rely on their expertise. This was especially evident in educational institutions, which had been radically reorganized, but still depended on the accumulated knowledge of incumbent faculty. The difficulties that these changes presented to members of the old educated elite were substantial, but none threatened to dislodge them from the upper echelons of society or greatly diminish their central position in the education realm. That would change in 1957, when the CCP embarked on a much more radical effort to level class differences in the cultural field.

The Education Revolution

By the late 1950s, Mao began to privately criticize contemporary Soviet policies, including its education policies, for being too conservative. His thinking was reflected in commentaries he wrote criticizing a Soviet economics textbook used in Chinese schools. Although he concurred with the Soviet author that once socialist production relations had been established there were no longer classes plotting to restore capitalism (he would later change his mind), he nevertheless insisted, "In a socialist society there are still conservative strata and something like 'vested interest groups.' There still remain differences between mental and manual labor, city and countryside, worker and peasant. Although these are not antagonistic contradictions they cannot be resolved without struggle."[13] Mao had become convinced that Soviet policies tended to preserve these class differences, and he began to push for more radical policies.

After Mao encouraged intellectuals to criticize the party's shortcomings in the spring of 1957, he also led the counterattack by party officials against intellectuals.[14] In the end, he not only gave his blessing to the subsequent Anti-Rightist campaign, but he transformed it into a much broader movement to undermine the influence and social position of the old educated elite. Under his direction, the Anti-Rightist movement gave way to the Great Leap Forward, a monumental undertaking designed to move China dramatically closer to the Communist society envisioned in Marxist doctrine. Although the Great Leap is remembered mainly for its program of rapid economic construction, it also encompassed an ambitious agenda in the cultural field, which at some moments was called a Cultural Revolution and at others an Education Revolution. This revolution was presented as a natural continuation of the Communist class-leveling program, extending it from the economic to the cultural field. Having eliminated the economic dominance of the old elite classes, the CCP could now turn its attention to their cultural dominance. The redistribution of cultural resources, however, would prove much more complicated than the redistribution of material wealth.

It was not difficult to win the support of many Communist cadres for a new round of attacks on the old educated elite. Party cadres, most of whom had relatively little education, were often in direct competition with members of the old elite in their workplaces (from central ministries down to factory workshops and primary schools), and their children were beginning to compete in the academic and political credentialing systems. Thus, the party cadres personally benefited from political campaigns that handicapped the old elite and undermined the cultural foundations of their power. Their motivations, however, were more than simply instrumental; deep moral convictions also inspired them to support class leveling in the cultural field. The party had cultivated in its members indignation at the social and political pretensions of the old elite classes, and Communists had internalized the idea that the unequal distribution of cultural resources—like the unequal distribution of private property—

was unjust. Party cadres were overwhelmingly of peasant origin, they identified with the peasantry, and they considered themselves to be the workers' and peasants' representatives.

The radical slogans and practices that convulsed Chinese schools during the Education Revolution that accompanied the Great Leap Forward were elements of a radical program to eliminate class differences in the cultural field. This program can be divided into three aspects—increasing political power at the expense of cultural power, redistributing cultural capital, and altering conventional academic credentials and occupational categories in order to combine mental and manual labor.

INCREASING POLITICAL POWER AT THE EXPENSE OF CULTURAL POWER

Communist authorities refused to cede any ground in response to the arguments raised by Qian Weichang and others in 1957 that nonexperts should not lead experts. Instead, they reiterated even more strongly their insistence that all other considerations—including academic, technical, and economic considerations—must be subordinated to the general political orientation established by the CCP. Other interests, they argued, were narrow and particularistic; only party deliberations could properly define the needs and interests of the broad masses of people. This view was expressed in the slogan, "politics takes command" (*zhengzhi gua shuai*), and a logical corollary was that experts should be subordinated to political leaders. At the outset of the Great Leap Forward in 1958, Mao exhorted Communists not to be intimidated by intellectuals: "Professors? We have been afraid of them ever since we came to the towns. . . . When confronted by people with piles of learning, we felt we were good for nothing. . . . I believe this is another example of the slave mentality. . . . We must not tolerate it any longer."[15]

Communist leaders were profoundly concerned that expertise continued to be almost entirely in the hands of nonparty members. Most experts were from the old elite classes and their allegiance to the Communist program was questionable. In most work units and localities, poorly educated Communist officials were attempting to supervise non-Communist experts; even though they could formally impose their authority, their ability to exercise leadership was compromised by their lack of technical understanding. To address this problem, the CCP provided technical training for its cadres and worked to win the political allegiance of incumbent experts. But the long-term solution was to train a new generation of cadres who had expertise and were also committed Communists, that is, cadres who were both Red and expert.

In the spring of 1958, the CCP launched a campaign to Replace White Flags with Red Flags. At Tsinghua, the school newspaper extolled university cadres and teachers deemed to be the university's Red flags, including young cadres who had demonstrated both outstanding political leadership and impressive technical expertise and senior faculty members who had recently joined the party. These new Red flags, the articles argued, were proving themselves superior in every way to the old White flags, who had taken so much pride in their academic and technical expertise. A central goal of the campaign was to "take schools out of the hands of bourgeois intellectuals." It not only consolidated the power of the university party organization, but it also brought changes in the daily practices of university departments. Previously, although most senior professors had been careful not to offend the party leadership and had deferred to them on questions that had clear political implications, many had insisted on taking charge of designing curricula. Now they were compelled to surrender this authority to the collective wisdom of the teaching and research groups, in which younger faculty members participated and in which political as well as academic considerations were weighed. Young teachers, especially those who had joined the party, were encouraged by party leaders to "revolutionize" the university. Yang Yutian, who graduated from Tsinghua and began teaching at the university in 1959, recalled, "When we became teachers the idea was that the older teachers had been educated in bourgeois society and that we younger teachers were trained under the proletarian society, so we had to take over and occupy the battleground."[16]

REDISTRIBUTING CULTURAL CAPITAL

While replacing White flags with Red flags was a major undertaking, it was not nearly as ambitious as the Communist goal of radically redistributing educational attainment across the population. This was to be accomplished by rapidly expanding the school system and by imposing class preferences in school admissions. Between 1958 and 1960, hundreds of thousands of primary schools and tens of thousands of junior middle schools were built in rural areas, and the proportion of children who attended schools at both levels increased significantly. The newly established rural communes were instructed to build village-based junior middle schools to provide practical knowledge and skills connected to rural life; this was the first large-scale introduction of secondary education at the village level. While the effort to expand access to education was concentrated mainly in villages and poor urban communities, Tsinghua was also drawn in. A night school was established to provide basic education as well as technical training for workers at the university. Over 1,000 workers participated in these courses and hundreds of other workers and family members attended literacy classes.[17] In 1958, the middle school attached to the university, which until then had exclusively served children of people employed by Tsinghua and neighboring Peking University, was instructed to "open its doors to workers and peasants," which in this case meant recruiting children from neighboring villages. As a result, about 20 percent of the students at the university's attached middle school were now recruited from village primary schools.

Tsinghua also developed special programs to prepare poorly educated working-class and peasant cadres to take administrative positions in economic enterprises and government offices. These programs had begun much earlier, with the establishment in 1952 of a "workers and peasants accelerated middle school" on the Tsinghua campus. The first cohorts of students were primarily peasants who had fought in the civil war, and later cohorts also included factory workers who had shown political promise; almost all of the students were party members. Tsinghua gave the school high priority. Zhou Peiyuan, the eminent physicist who had studied under Albert Einstein, was asked to serve as the school's principal, and especially talented and motivated teachers and cadres were assigned to the program. Altogether, 903 students graduated from the three-year accelerated program during the 1950s. Of these, 567 continued on to college (including 147 who attended Tsinghua University), and many of them went on to serve in important leadership positions. The school, which was designed to help train cadres in the period of regime transition, was closed in 1958.[18] By that time, Tsinghua had established larger short-term training programs for Communist cadres who had already begun working in administrative positions.

Although the CCP had long discriminated based on family origin, class line policies came to the fore during the Great Leap Forward. Before then, relatively few university students were of working-class or peasant origin because the great majority of those who were able to pass the entrance examinations were from educated families. As of 1956, only one in five students enrolled in Tsinghua was of worker-peasant origin (a category that included the children of revolutionary cadres).[19] Wei Xuecheng, the son of a rich peasant who entered the university in 1953, recalled that school officials did not pay much attention to class designations in the early 1950s, "because everyone's family origin was bad."[20] By the late 1950s, however, an increasing supply of working-class and peasant origin middle school graduates coincided with the new political offensive against the old educated elite. "In 1958, they started to look more at family origin," recalled Yang Yutian, a student of poor peasant origin who entered Tsinghua in 1956. "The thinking then was, 'We need to train more poor and lower-middle peasants' kids, and not so many landlords' and capitalists' kids.'"[21]

Class line preferences were explicitly used to counter advantages enjoyed by children of the old elite in the examination competition. In 1958, Tsinghua suddenly granted special admission to several hundred young soldiers and graduates of workers and peasants accelerated middle schools (including the one run by the university). Despite their lower level of academic preparation, these students were integrated into the same classes as students who had passed the exams. The academic disparity between the two groups caused teaching difficulties, but in the ideological environment of the Great Leap Forward, other students and teachers were solicitous of these "worker-peasant students." Many of them were already party members, which won them esteem among their classmates and leadership positions in student organizations. Moreover, during the Great Leap Forward, teachers were encouraged to pay more attention to students who were having difficulties, rather than to those who were excelling. This was part of the general push at that time to popularize education, rather than concentrate on elite training, an orientation that was accompanied by an ideological campaign to promote the "mass road" as opposed to the "genius road" in education. The former, it was argued, not only promoted equality, but would lead to faster national development by bringing out the full abilities of the masses of workers and peasants, while the latter both perpetuated inequality and hindered national development.

COMBINING EDUCATION AND PRODUCTIVE LABOR

According to a key slogan of the Great Leap Forward, education was henceforth to "serve proletarian politics and be combined with productive labor." In adopting the Soviet model in the 1950s, Chinese universities had already greatly reoriented their curricula toward production and practical training, and Tsinghua and other engineering schools had adopted a Soviet requirement that students carry out semester-long graduation projects that simulated real-life problems. The practical orientation of the Soviet university system reflected an effort to reduce the distinctions between abstract learning and practical work, inspired by the Communist ideal of eliminating the differences between mental and manual labor. During the Great Leap Forward, Chinese universities moved a great deal farther in this direction, in two respects.

First, teaching methods and materials were revised to further emphasize the combination of theory and practice. Now Tsinghua students were required to take on "real knives and guns" graduation projects, including participating in the design and construction of major reservoirs and other public works projects, and in the design and manufacture of machinery and equipment. Tsinghua also sent small groups of teachers and students to forty-five counties and 150 work units to help with industrial projects, including ill-fated efforts to build small-scale iron and steel furnaces. In addition, the university built sixty-one factories and workshops on or near its sprawling campus. These factories physically transformed the university, changed the composition of its workforce (which now included over two thousand industrial workers), and substantially altered the way education was conducted. The Soviets had carefully distinguished between institutions that specialized in higher education (universities), research (research institutes), and production (factories); the Chinese now made a point of combining all three, making campus factories into centers of teaching, research, and production. Tsinghua's success in these industrial endeavors established the university's position as the country's premier school of technology and Jiang Nanxiang's reputation as a pioneer of a specifically Chinese road in higher education.

Second, students and teachers were now required to participate regularly in manual labor. This practice began in the fall of 1957, as students and teachers went to help harvest crops in nearby communes, and continued on a much grander scale in the spring of 1958, as 6,500 Tsinghua students and teachers were dispatched to take part in the construction of a huge reservoir on the outskirts of Beijing, moving earth with wheelbarrows. Over the next two years, Tsinghua students and teachers continued to harvest crops in nearby communes, participated in numerous construction projects at the university and in the surrounding area, and worked regularly in the new campus factories.

In required Marxist-Leninist philosophy classes, Tsinghua students had learned that the division between mental and manual labor would eventually disappear with the advent of communism. For nearly a decade after the CCP came to power, however, students could regard this as a distant goal that had limited practical bearing on the present, in which their lives were consumed with learning engineering formulas required for the professional, technical, and administrative jobs awaiting them after graduation. Then suddenly, during the Great Leap Forward, radical egalitarian principles appeared to be reshaping the whole world. Mao submitted that the goal of universities should be to train "laborers with socialist consciousness and culture." Articles in Tsinghua's school newspaper suggested that soon the differences between mental and manual occupations would be reduced to a minimum, and students should embrace the goal of becoming a "common laboring person."

In the fall of 1958, while Tsinghua students were building dams and harvesting crops, they were also attending meetings to debate the implications of this new vision of university education. The school newspaper was filled with detailed accounts of these meetings and with students' essays debating philosophical questions posed in the meetings. Was it necessary to have a system of ranks and to distribute material goods unequally in order to motivate people? Was there utility in the division between mental and manual labor? Should highly educated university students and professors spend their time hauling earth? Should students be proud to be attending an elite institution like Tsinghua University? Did individual academic accomplishment reflect inborn talent or differences in social environment? Could striving for individual fame serve a positive social purpose? Opposing essays presented the strongest possible arguments on both sides, while leaving little doubt as to what the correct Communist view was.

Revival of Meritocracy

Amid the wreckage left by the collapse of the Great Leap Forward in 1960 were the ambitious goals of the Education Revolution. As the economy staggered to a crawl and famine enveloped large swaths of rural China, tens of thousands of new schools were shuttered and enrollment at all levels was sharply cut. At Tsinghua, the frenetic activity of the Great Leap gave way to listlessness, as the university party organization struggled to feed the campus population, and physical activity was cut to a minimum to preserve energy. The collapse of the Great Leap resulted in the curtailment not only of plans to rapidly expand the school system, but also of radical policies designed to alter the character of education.

In 1960, as a result of Tsinghua's prominence in the recent efforts to establish a distinctly Chinese road for higher education, Jiang Nanxiang was named deputy minister of education with responsibility for higher education.[22] In this capacity, he played a critical role in organizing the retreat from the Great Leap and setting the new path forward, a path that was guided by the motto, "Readjust, consolidate, replenish, and raise the level." In 1961, Jiang oversaw the drafting of a document containing sixty articles that served as the principal policy guidelines for higher education during the next five years, until the Cultural Revolution.[23] The document, popularly known as the "Sixty Articles," set the tone for a period of retrenchment and recovery that was marked by a retreat to conventional educational practices and a sharp turn away from popularization in favor of shoring up elite education, an orientation that allowed educators to embrace meritocratic ideals. Jiang and other education officials concluded that the Great Leap had brought a number of distortions in the field of education, of which they were particularly concerned about two: an excessive stress on labor at the expense of teaching, and the fostering of an egalitarian mentality that undermined quality.

Tsinghua and other technical schools retained the approach of combining teaching, research, and production that had been pioneered during the Great Leap Forward, but emphasized teaching and greatly scaled back campus-based production. Scores of small factories and workshops that Tsinghua had established were closed, and hundreds of workers (many of whom were demobilized soldiers) were sent to their home villages. Nevertheless, several major facilities were retained, including an experimental nuclear reactor and factories that produced advanced machine tools, industrial equipment, and computers. Jiang continued to stress the importance of hands-on learning, sending students to gain practical experience in these factories as well as others outside the university, but the emphasis was on technique and he had little use for manual labor in the curricula.

After the meritocratic turn, Tsinghua only accepted students with examination scores that surpassed the school's very high threshold, but it continued to employ class preferences. Teachers assigned to recruit the school's quota of students from each province would examine the application files of top-scoring candidates, considering reports summarizing their family origin and political performance. Those with problematic family histories faced exclusion, although extraordinary examination scores or political performance might compensate. Family background continued to be important in determining assignment to academic departments, and it was difficult for students of dubious origin to enter military-related fields, such as chemical engineering, nuclear physics, radio engineering, instrumentation, and other electrical engineering disciplines.

Differences in quality among schools were formalized and reinforced by the establishment of a "keypoint" system, which funneled greater resources to select schools.[24] At each level, the entrance examinations systematically divided successful candidates into schools of different qualities. Education officials named keypoint universities at the national and provincial levels, and keypoint middle schools at the provincial, county, and district levels. Even primary schools were divided into keypoint and ordinary categories. In line with this policy, in 1960 the middle school attached to Tsinghua University was converted into a keypoint for the Beijing municipal district. Until then, the school had served the children of all employees of Tsinghua and nearby Peking University, and the disparity in educational level between the children of professors and ordinary university workers created problems for teachers, a problem that was exacerbated during the Great Leap Forward by the admission of village children. After the school was converted into a citywide keypoint, it recruited exclusively from among the city's top-scoring students, eliminating the difficulties created by uneven abilities. Most of the professors' children were able to test into the middle school, but the university workers' children and the village children had little chance to get in. In its new incarnation as a keypoint, Tsinghua's attached middle school was expanded by adding senior middle school grades and was relocated to a large and beautifully appointed new campus on land appropriated from a village just north of the university. It quickly became one of Beijing's best middle schools, and in the early 1960s, about 85 percent of its graduates tested into college, a remarkable proportion at a time when only about 1 percent of young people attended college. Many students chose to attend the middle school with the intention of going on to Tsinghua University, and between 1960 and 1966, 139 of its graduates achieved this goal.[25]

As the deputy minister of education, Jiang Nanxiang led a campaign to oppose "egalitarianism." As part of this campaign, in 1961 Tsinghua University Deputy Party Secretary Liu Bing argued that school officials and teachers had to give up egalitarian thinking in order to inspire young people to achieve their full potential. "We should encourage young people's ideals and ambitions," he declared. "Criticizing the idea of becoming an expert and becoming famous will hurt young people's enthusiasm—young people ought to become experts and become famous to serve the people."[26] In 1962, Tsinghua inaugurated a program designed to "teach according to ability" (*yincai shijiao*). The students selected to participate (about 3 percent of the student body) were moved to a separate dormitory with better facilities, and the university's best professors were assigned to teach them, offering individualized instruction. The program specifically overturned Soviet-inspired prohibitions against dividing students by ability, which Jiang now declared had caused the university to fail to pay proper attention to its best students.[27] In an article in the teachers' edition of the Tsinghua University newspaper, Professor Jin Xiwu argued that egalitarian thinking had to be eliminated in order to identify and cultivate special talent. "It is not enough to generally oppose egalitarianism in teaching; we must especially pay attention to the problem of cultivating the minority of superior students. We must recognize that individuals' intelligence and abilities are different. We can try to lessen the difference, but efforts to do so have limits. Just as everyone can't climb Mount Everest, everyone can't scale the heights of science."[28]

The same principle was applied at the primary and middle school levels. At Tsinghua's attached primary school, a group of first graders was selected to participate in an "experimental class" with an accelerated curriculum. The participants were all professors' children who had attended the university's preschool program (workers generally did not send their children to the preschool program, which charged tuition). At Tsinghua's attached middle school, entering students were divided into what were informally called "fast," "medium," and "slow" classes based on math and foreign language tests, and in 1964, the middle school initiated a special college preparatory program for fifty carefully selected senior middle school students. Classes were taught by the school's best teachers as well as by university professors, and science labs were conducted at university facilities. Plans were laid to admit the best graduates directly into Tsinghua without taking the college entrance examination (in order to promote creative thinking). The program was the brainchild of President Jiang, who was intent on developing Tsinghua's attached primary and secondary schools so the university could cultivate high-quality college recruits from within its own system.

Academic discipline at the middle school was reinforced by the demands of the college entrance examinations. Exam content determined middle school curricula and exam preparation consumed students' lives. All middle school students, including those whose parents lived nearby on the Tsinghua campus, were housed together in dormitories and their daily routine was highly organized. Before and after classes, they ate together and participated in collective exercises and sports, and in the evening they studied together under the guidance of teachers. Students lauded the dedication of the principal, Wan Bangru, and of the bright young teachers he recruited, many of whom put off getting married in order to concentrate on building the school's academic reputation.

During the early 1960s, both Tsinghua University and its attached middle school celebrated intense academic competition. Students competed—for individual and school glory—in citywide, provincial, and national academic and athletic competitions. The ascendance of meritocratic ideas led to a dramatic change in teachers' attitudes toward students. "Between 1958 and 1960, teachers were more concerned about helping the students whose study was weaker; they didn't want them to fall behind—the thinking then was you can't allow a single student to fail," recalled Yang Yutian, a young university teacher. "But after 1960, the thinking changed—teachers should pay more attention to the better students."[29] There was no longer room for the worker-peasant students who had been admitted to Tsinghua in 1958 without passing the entrance examinations; they were put in separate classes and then graduated early.

TWO TRACKS: POPULARIZATION AND ELITE TRAINING

As China recovered from the collapse of the Great Leap in the early 1960s, Mao moved to bring back the party's class-leveling agenda, reminding his followers to "never forget class struggle." In remarks delivered during the Spring Festival of 1964, he once again called for a "revolution in education." He expressed disdain for conventional teaching practices, encouraging students to whisper to each other during exams and suggesting they sleep in class if they found lectures boring.[30] Education officials were bewildered and disturbed; Jiang Nanxiang famously told other leaders at Tsinghua that the speech would have to be "translated a little" for engineering students.[31] Mao's fickle words, however, were soon followed by concrete initiatives that were less difficult to decipher. Radical education slogans and practices from the Great Leap Forward were revived along with a crash program to build rural schools. With the help of Kang Sheng and other proponents of radical education policies, Mao sent work teams to Peking University and other schools to investigate education practices.

Tsinghua responded to this new radical "wind" by cutting the length of its undergraduate program from six to five years, reducing the number of hours students spent in classes, experimenting with open-book exams, encouraging students to express divergent opinions in class, emphasizing the practical orientation of coursework and increasing the time students spent in the field, stepping up political and military training, mobilizing students and teachers to help nearby communes at harvest time, and emphasizing class line policies in student recruitment. The university also started "half-work/half-study" industrial and agricultural technology programs. Junior middle school graduates from the Beijing area were recruited for the industrial technology program and they also worked part-time in the university's factories. Young villagers were selected by nearby communes to take part in the agricultural technology program and they were expected to return to their communes after graduation.[32]

These half-work/half-study programs were one way education officials sought to accommodate Mao's pressure to popularize education, while also maintaining high standards at elite institutions. The aim was to establish a clear division between popular and college preparatory education. This approach, championed by President Liu Shaoqi, was known as "two kinds of education systems, two kinds of school systems." The aim was to protect elite education from dilution, while at the same time protecting vocational education from the seductive pull of examination preparation.[33] In 1964, Jiang Nanxiang concisely summed up this two-track philosophy. "In our country's current education situation, from a certain perspective you could say that 'the high is not high enough' and 'the low is not low enough,' so neither has been accomplished," he argued. "'The high is not high enough' means that currently our country's universities are not only behind the universities in the advanced industrialized countries in scientific skill level and key equipment, but there continues to be a huge gap. . . . 'The low is not low enough' means our regular and higher level education both lay heavy stress on formal education and don't pay enough attention to half-work/half-study, half-plough/half-study, and the other forms of simple, part time, and correspondence schools."[34]

This type of two-track philosophy had originated during the Great Leap Forward, when the initial plans for both highly selective keypoint academic schools and part-time agricultural middle schools were adopted. During those heady days, when all goals were to be accomplished simultaneously and quickly, Mao had endorsed this approach. During the more sober days of the mid-1960s, however, he began to see this two-track approach as a recipe for entrenching class differences. Mao, therefore, moved in precisely the opposite direction. Instead of raising the top and lowering the bottom, he resolved to make the entire education system level, so that all young people would attend schools of equivalent quality for the same number of years and then go to work. In 1966, he would launch a more thoroughgoing assault on the cultural foundations of class power, and the first step would be the elimination of the examination system.

Chapter Three

Cradle of Red Engineers

On March 21, 1958, in the first days of the Great Leap Forward, Jiang Nanxiang addressed thousands of students and teachers gathered for an all-school meeting. "Our requirement for training cadres," he told them, "is that they become both Red and expert. If you are not both thoroughly Red and deeply expert, that is a huge waste."[1] During the coming months, students at Tsinghua, like those in other universities across the country, were called upon to join a "great debate" about the Red and expert question, expressing their opinions in big character posters and at gatherings of their classmates. Students had previously been encouraged to "take the Red and expert road," but this campaign made clear that from now on they would face much more rigorous political expectations. It would no longer be possible to concentrate on academic study and leave politics for others, an approach that was stigmatized as "taking the White expert road." Everyone was now expected to combine diligence in academic studies with political activism under the guidance of the Youth League and the party. Tsinghua was to become a "cradle of Red engineers."

The initial impulse behind the Red and expert drive was class warfare. The new regime distrusted the old "White experts" and it was determined to replace them with a new generation of cadres who not only had expertise, but were also committed to the Communist project. As part of the effort to replace White experts with Red experts, the CCP organized crash programs to educate worker-peasant cadres, built hundreds of thousands of new schools in villages and poor urban areas, created literacy and technical-training programs in factories, and implemented class line policies that discriminated against children of the old elite in order to make room for children of workers, peasants, and revolutionary cadres. These policies were implemented with a new sense of urgency following the Anti-Rightist campaign of 1957 and in 1958 they were animated by the revolutionary enthusiasm of the Great Leap Forward.

In the long run, however, the CCP's determination to combine Redness and expertise became a formula for class conciliation. It ended up promoting elite convergence by encouraging the new political elite to accumulate cultural capital and the old educated elite to accumulate political capital. As members of both the old and new elites—and their children—strived to become Red experts, a new social stratum emerged whose members had secured similar credentials, acquiring a stake in both the political and cultural pillars of the emergent class order.

Reproduction and Convergence of Elites

RED EXPERTS AT THE INTERSECTION OF TWO ELITE WORLDS

In 1949, at the beginning of the Communist era, the new and old elites made up exceedingly small portions of the Chinese population. One way to gauge the size of both groups is to use the class categories employed by the new regime. The figures presented in Table 3.1 estimate the distribution of the population across these categories. Membership in these classes was based on the status of the family head during the period immediately before the 1949 Revolution. Three of the categories directly corresponded to the new political elite: revolutionary cadres, revolutionary soldiers, and descendants of revolutionary martyrs; together they made up slightly less than 1 percent of the population. Two categories—landlords and rich peasants—can be described as old rural elites, and three categories—capitalists, white-collar employees, and independent professionals—can be described as old urban elites. The old rural elite categories made up 4.3 percent and the old urban elite categories made up 1.4 percent of the country's population.

A sense of the size of the old and new elites can also be conveyed by the number of people who were party members or held a college degree. In 1949, out of an adult population of nearly 400,000,000, there were less than 5,000,000 party members and less than 185,000 college graduates.[2] Potential members of

TABLE 3.1
Class origin categories as proportions of the population

"Class line" status	Category	Percent of survey respondents
Laboring	Revolutionary cadre, soldier, or martyr	0.9
	Worker or poor/ordinary urban resident	5.1
	Poor or lower-middle peasant	76.8
Other	White-collar or independent professional	1.2
	Upper-middle peasant or small entrepreneur	11.5
Exploiting	Capitalist	0.2
	Landlord	2.3
	Rich peasant	2.0

NOTE: Data in this table were derived from a national probability sample survey of over six thousand urban and rural residents conducted in 1996. Respondents were asked to select one of fifteen family origin categories. The categories included three—"Rightist," "Bad element," and "Counterrevolutionary"—that were used to designate political status, together with twelve categories that were used to designate class status. Because almost all Rightists were intellectuals, I combined the two individuals in the "Rightist" category with the "White-collar or independent professional" category. I placed a handful of individuals who selected the "Bad element" or "Counterrevolutionary" categories or who failed to select any of the categories into a residual group. I included only respondents who were born before 1957 because individuals who came of age in the post-Mao era, after the family origin system was abandoned, were not as likely to be aware of their family's designation. The survey is described in Treiman et al. (1998).

the new political elite also included several million soldiers and revolutionary activists who were not yet party members but would later be recruited and appointed to official positions. Among the old educated elite we might also include those who had obtained a senior middle school diploma, which at that time was a rare educational accomplishment. Even if senior middle school graduates are counted, however, the cultural elite made up less than 1 percent of the population. The overwhelming majority were illiterate or had attended only a few years of primary school.[3]

If each of the two elites was tiny, the number of people who could claim to be part of both groups—intellectuals who had joined the Communist Party—was infinitesimal. Most of this group had been recruited as students by underground party organizations in middle schools and universities. The importance of this group, of course, was far greater than its size, as many of these students rose to positions of leadership in the movement. Indeed, many of the very top Communist leaders were of elite origin and were well educated. Of the forty-four members of the Seventh Central Committee, which was elected in Yan'an in 1945, at least 27 percent were of upper-class origin and at least half had studied abroad in advanced capitalist countries.[4] Tsinghua University became an important recruiting base for the CCP especially during two periods of political upheaval—the December 9 (1935) student movement to protest Japanese aggression and the civil war years immediately preceding the Communist victory in 1949. During these two periods, hundreds of Tsinghua students joined the underground Communist movement and several of them went on to become high-level party officials.[5] Educated Communists, however, made up only a small fraction of the millions of party cadres, the great majority of whom were poorly educated peasants. In Heilongjiang Province, for instance, 95 percent of provincial party members were of poor peasant, hired hand, or worker family origin. Nearly three-quarters were functionally illiterate, most of the rest had only attended primary school, and only 2.4 percent had been to middle school.[6] It is possible that the educational level of party members in other provinces was higher, but throughout the country the party was made up largely of rural recruits from humble origins, and it had relied on these peasant cadres to govern its vast rural base areas.

Before 1949, the tiny handful of Communist intellectuals linked two worlds that were geographically distant and were at located at opposite ends of the social hierarchy. The Communist movement was based in remote rural base areas, while the educated elite were increasingly concentrated in China's more prosperous cities. The occupants of both of these worlds—the Communist cadres and the educated urban classes—were highly self-conscious elites with very distinct credentials, cultures, and value systems. When the Communist Party took over China's urban institutions and the two worlds collided, educated Communists occupied a tiny and highly distinctive social space at the intersection of both groups. Over time, however, the population of this intersection grew steadily, augmented by the Red experts produced in China's elite schools. Most of these Red experts had their origins in one or the other of the elite groups, but a growing number were also recruited from below.

CREDENTIALING SYSTEMS AS MECHANISMS OF ELITE REPRODUCTION AND CONVERGENCE

The academic and political credentialing systems described in the previous two chapters regulated access to cadre positions as experts and administrators. Using Bourdieu's language, these credentialing systems distributed cultural and political capital, and they provided mechanisms for elite families to reproduce their advantageous status. This was most readily accomplished *within* the cultural or political field, but these institutions also facilitated what Bourdieu called "capital conversion," that is, using capital accumulated in one field to obtain access to capital in another field.[7] In postrevolutionary China, the most important capital conversion strategies spanned the political and cultural fields.

Of course, the prescribed purpose of the academic and political credentialing systems was to select and train qualified candidates to serve as party and state cadres, not to promote elite reproduction. Neither system excluded non-elites. On the contrary, class line policies gave preference to children of workers and peasants, and the massive expansion of primary and middle schools was designed to provide more opportunities for children from modest homes so as to disperse cultural capital. Nevertheless, both systems operated in a fashion that facilitated elite reproduction and children of the new political and the old educated elites were much more likely to win credentials than were children of workers and peasants.

Elite families—both old and new—could only retain their advantageous social positions across generations by means of strategy and diligence. Their children had to succeed in arduous competitions for academic or political credentials; and success was by no means guaranteed because both credentialing systems required individual achievement. Well-endowed parents, however, could provide substantial advantages. Revolutionary cadres could pass on to their children pride in their revolutionary heritage and motivation to follow in their parents' footsteps, as well as political connections and knowledge about how to get ahead in the political recruitment system. Intellectuals could pass on to their children similar advantages in relation to the education system (pride, motivation, social connections, and knowledge about how to get ahead in school), and could also convey knowledge invaluable in the academic competition. Thus, family inheritance provided privileged access to political capital for children of the political elite and privileged access to cultural capital for children of the cultural elite. In less direct ways, political capital also provided privileged access to cultural capital, and cultural capital provided privileged access to political capital. The two credentialing systems were intricately intertwined, and the links between them became mechanisms for capital conversion.

On the one hand, possession of political capital provided several advantages in winning cultural credentials. First, one of the required college entrance examinations specifically assessed mastery of political doctrine. Second, in considering the applications of candidates who passed the entrance examinations, college recruiters took into account reports compiled by middle school teachers that summed up an applicant's political performance and family origin. Third, Communist authorities established a variety of special schools in which admission was based largely on political qualifications. These included schools for adult party members and cadres, such as the worker-peasant accelerated middle school and the special cadre training programs created by Tsinghua University in the 1950s, as well as primary and middle schools that catered to children of Communist officials and military officers. The latter were an outgrowth of primitive boarding schools created to care for children of revolutionaries during the Anti-Japanese War and the Civil War. After 1949, these schools were transformed into highly elite institutions; some groomed students for careers in the military, while others prepared students for the university examinations. In the mid-1950s, as the exclusive nature of these schools began to smell of aristocratic privilege, they were required to open their doors to noncadre children, but they continued to cater largely to children of the political elite.

On the other hand, possession of cultural capital also provided advantages in winning political credentials. The most important advantage was that elite schools became a focal point of political recruitment. Moreover, recruitment into the Young Pioneers, the Communist Youth League, and the Communist Party was based on not only political, but also cultural criteria. Academic qualifications were particularly important in middle school, where teachers used the prize of Youth League membership to inspire greater discipline in preparing for the college entrance examinations. The political recruitment apparatus at universities placed more emphasis on political criteria (especially when it came to joining the party), but students could not expect to go far in the political competition if they fell behind in their studies. This point was continually reiterated by education authorities. "Examination grades are not only the standard for measuring the degree of academic accomplishment," declared a 1962 Ministry of Education document drafted under Jiang Nanxiang's direction. "Through grades we can also discern the student's political character."[8] Moreover, many of the political criteria used to evaluate potential Youth League and party recruits required cultural proficiency. Mastery of the nuances of political doctrine, articulate elocution, writing skills, fine calligraphy, musical talent, and theatrical abilities were all highly valued in political activism. Students who excelled in language and humanities classes, therefore, often did well as political activists.

The academic and political credentialing systems, thus, not only fostered the reproduction of the political and cultural elites, but also enabled the gradual convergence of the two by providing mechanisms for capital conversion. Both the reproduction and convergence of elites can be seen by examining the class origins of faculty and students at Tsinghua.

REPRODUCTION AND CONVERGENCE AT TSINGHUA

In 1970, Tsinghua University conducted an investigation into the family histories of its teaching staff, producing the data presented in Table 3.2. By comparing the class origins of senior and junior faculty members, we can gauge the extent to which policies designed to redistribute cultural capital during the first two decades of Communist power had changed the social composition of this segment of the educated elite. After 1961, the university virtually stopped promoting faculty members to the ranks of associate and full professor, and almost all teachers who had achieved these ranks had been hired before 1949. Most lecturers had been hired in the 1950s, and most assistant teachers (the starting position of new faculty members) had been hired in the 1960s.

TABLE 3.2
Class origins of Tsinghua University faculty, 1970

	Full and associate professors		Lecturers		Assistant teachers		Totals	
	No.	%	No.	%	No.	%	No.	%
Worker and poor or lower-middle peasant	3	1.5	43	4.5	216	15.0	262	10.2
White-collar/ middle peasant	52	27.4	301	31.6	680	47.3	1,033	40.0
Exploiting	114	60.0	485	50.9	419	29.1	1,018	39.5
Other	21	11.1	123	12.9	123	8.6	267	10.3
Totals	190	100.0	952	100.0	1,438	100.0	2,580	100.0

SOURCE: Tsinghua University (1975).

Tsinghua's senior faculty members (associate and full professors) were virtually all from the old elite classes. Over 60 percent came from families determined to be part of the exploiting classes, which combined, made up less than 5 percent of the population. Over 27 percent were counted in the middle category, which included white-collar employees and senior middle peasants (the great majority of these, it can reasonably be assumed, were from white-collar—that is, educated elite—families). These figures show the extent to which possession of cultural and economic capital was correlated in Chinese society before the Communist era, and they indicate that children of working-class and peasant origin had been almost completely excluded from the most elite academic institutions. While workers and poor and lower-middle peasants made up over 80 percent of China's population, only 1.5 percent of the senior faculty originated from these groups.

The impact of Communist redistribution efforts can be seen in the composition of the junior faculty. Class line policies clearly limited the opportunities available to those encumbered by exploiting-class labels: their proportion dropped from 60 percent of the senior faculty, to 51 percent of the lecturers, and 29 percent of the assistant teachers. Those that benefited the most from this discrimination against the "bad" classes were children of the middling classes, whose proportions grew from 27 percent of the senior faculty, to 32 percent of the lecturers, and 47 percent of the assistant teachers. Most of these new faculty members were undoubtedly from urban educated families, but many were also from middle peasant families, who were in the best position to take advantage of the expansion of middle schools in rural China (where discrimination against old elite families was particularly acute). The numbers of junior faculty members of working-class and poor and lower-middle peasant origin were also increasing steadily, but the proportions were still small: only about 5 percent of lecturers and 15 percent of assistant teachers. Thus, even among the most recently hired cohorts of faculty members, the overwhelming majority were from old educated elite families. Despite significant Communist redistribution efforts, the composition of the Tsinghua junior faculty indicates the extent to which the highly skewed distribution of cultural capital was being reproduced.

Redistribution policies had a greater impact on the composition of the Tsinghua University student body, as can be seen in the figures presented in Table 3.3. The proportion of students of worker-peasant origin—a category that by contemporary convention included not only students from worker and

TABLE 3.3
Worker-peasant proportion of student enrollment, Tsinghua University, 1952–1964

	1952	1953	1954	1955	1956	1958	1959	1964
Total enrollment	3,269	4,234	5,214	6,329	8,647	10,889	11,366	10,771
Worker-peasant proportion	14%	17%	19%	18%	21%	32%	36%	44%

SOURCES: *New Tsinghua* Editorial Committee (1957, 120); *QHDXYL* (1959, 1960); *JGS* (Oct. 12, 1967).

poor and lower-middle peasant families, but also those from revolutionary cadre families—grew steadily during the 1950s, then jumped during the Great Leap Forward, and continued to rise impressively during the early 1960s, reaching 44 percent by 1964. Some of these students would eventually be hired to work at the university, but the low proportion of faculty members of worker-peasant origin suggests that these students were more likely to be assigned to nonacademic positions in industrial enterprises and government ministries.

The figures presented in Table 3.4 offer a more detailed picture of the composition of the student bodies at Tsinghua University and at its attached middle school during the early 1960s. The figures come from two sources. One column for each school records statistics, presumably culled from school records, published in student factional publications during the Cultural Revolution (the schools maintained—but did not publish—detailed breakdowns of the class compositions of their student bodies). A second column for each school records the recollections of interviewees who attended the schools between 1960 and 1966 about the family origin of their classmates. The interviewees were not selected at random and so they cannot be considered a representative sample, but their compiled recollections yielded figures that roughly correspond with those published in Red Guard newspapers.

According to students who attended Tsinghua's attached middle school, nearly two-thirds of their classmates came from old elite families (white-collar, independent professionals, or exploiting classes), a proportion that is corroborated by the statistical data. This proportion was particularly high because Tsinghua's middle school gave preference to graduates of the university's primary school—many of whom were faculty children—by requiring higher examina-

TABLE 3.4
Students' class origin: Tsinghua University and attached middle school, early 1960s

	Tsinghua University Attached Middle School		Tsinghua University	
Family origin[1]	Statistics from Red Guard publications[2]	Students' estimates	Statistics from Red Guard publications[3]	Students' estimates
Revolutionary cadre	25%	25%		9%
Worker	9%	10%		12%
Peasant	1%	1%	44%	26%
White-collar or independent professional	52%	57%	46%	42%
Exploiting classes[4]	14%	7%	10%	12%

1. Nine interviewees who attended Tsinghua's attached middle school and fourteen interviewees who attended Tsinghua University classified their classmates by family origin. There were nearly four hundred students in the classes described by the middle school students and just over 430 students in the classes described by the university students. Both were boarding schools and members of classes (twenty-five to fifty students) lived and studied together from admission to graduation, allowing students to get to know their classmates well. Family origin became a particularly salient personal characteristic before and during the Cultural Revolution. Nevertheless, the sharpness of former students' memories varied considerably. Two students were able to recall all of their classmates' names and student numbers and provided many details about their family backgrounds. Others only provided rough estimates of the number of students in each class category. I did not include estimates that were very vague or incomplete.

2. These figures are for one class (about fifty students) at Tsinghua's senior middle school (Beijing Forestry Institute Red Guard Fighting Groups 1966).

3. A student factional newspaper at Tsinghua reported that 44 percent of the university's students were of worker-peasant origin and 10 percent were of exploiting class origin as of 1964 (*JGS*, Oct. 12, 1967). The former figure presumably includes children of revolutionary cadres (per contemporary terminology). The remaining 46 percent of students were presumably from families designated as white-collar employees, independent professionals, and other middling categories.

4. I have included in this category classmates from families suspect for political reasons (e.g., "counterrevolutionaries" and "Rightists").

tion scores for students admitted from other city schools. The main reason that children from intellectual families were successful in gaining admission, however, was that they were highly proficient in the examination competition.

About a quarter of the middle school's students came from families of revolutionary cadres, soldiers, and martyrs, a reflection of how the academic credentialing system was facilitating elite convergence.[9] This was mainly due to the fact that many revolutionary cadres' children did very well on the examinations. Children of both educated and political elite families had privileged access to the best primary schools because they lived in neighborhoods where these schools were located, or because these schools had links to work units—such as universities and government offices—in which their parents worked, or because their parents had connections with teachers and school officials. Altogether, children of these old and new elites made up about 90 percent of the students at Tsinghua's middle school, while only about 10 percent were workers' and peasants' children.

The high proportions of students from old and new elite families was in part due to the fact that the middle school was attached to a university and was located in the nation's capital, where government ministries, research institutes, and a disproportionate share of China's universities were located. Studies that have examined the student composition of keypoint middle schools in other cities, however, have also shown that the great majority of these students came from either intellectual or revolutionary cadre families.[10] These urban keypoint schools trained a large proportion of the students who won coveted places in China's few universities, and they were a key node in a system that reproduced class advantages.

The student body at Tsinghua University was considerably more diverse than that of its attached middle school; in fact, a substantial minority of the university's students were of peasant and working-class origin. The university recruited from the entire country, and by the 1960s the new regime had created a huge network of primary and middle schools in rural and poor urban areas. Although these schools were generally inferior, they had begun to produce large numbers of graduates of working-class and peasant origin, some who were able to test into county, municipal, and provincial keypoint middle schools. My interviewees indicated that nearly 38 percent of their classmates came from ordinary working-class and peasant homes. Many of these, including quite a few students I interviewed, came from poor rural families and had attended village primary schools.

Despite the increasing number of students of humble origin, however, the majority of Tsinghua University students continued to come from elite homes. Students estimated that 42 percent of their classmates came from white-collar and independent professional families, and 12 percent came from exploiting-class families, proportions that were roughly corroborated by university statistics. At the same time, according to interviewees' recollections, about 10 percent of their classmates were revolutionary cadres' children. There were a smaller proportion of revolutionary cadres' children at the university than at its attached middle school, not only because the university recruited from the entire country (instead of just the capital), but also because many Communist cadres had only started families after 1949 and their children had not yet reached college age. Nevertheless, revolutionary cadres' children were greatly overrepresented at the university, as were the children of the old educated elite. Together, these two groups made up a large majority of the student body, an indication of the extent to which the academic credentialing system was facilitating both cultural reproduction and elite convergence.

BUILDING A "PROFESSORS' PARTY"

Even though the party's efforts to change the social origin of Tsinghua's teaching staff had only met with limited success, it was able to steadily change the faculty's political complexion. Jiang Nanxiang worked diligently to increase party membership among the Tsinghua faculty, inspired by a vision that explicitly sought elite convergence. To realize this vision, two difficult tasks had to be accomplished: first, members of the old educated elite had to be compelled to accept party leadership and embrace the Communist program, and second, resistance within the party to recruiting members of the old elites had to be overcome. In the mid-1950s, the party launched a campaign to recruit "high intellectuals" (*gaoji zhishifenzi*) and Jiang became one of the campaign's leading advocates. He criticized the exclusionary attitude of many party members, who were skeptical about this policy, which they saw as unseemly conciliation with the old elite. "We don't trust high intellectuals enough and we have too low an estimate of their progress," he declared. "This is particularly evident in the closed-door mentality towards high intellectuals who apply to join the party. There are quite a few people who have progressive thinking and who meet all the conditions for joining the party, and yet they have not been recruited. . . . Six years after liberation, among all the professors and associate professors in Beijing's twenty-six universities, only seventeen have been recruited; among the first grade engineers in Beijing, only one has been recruited; among the top medical professionals in Beijing, not even one has been recruited. No wonder some people say, 'Eminent people revolve around the party, but can't enter, just like the moon circles around the earth without ever being able to touch it.'"[11]

In 1955, with great fanfare, Jiang invited Liu Xianzhou, a highly respected Tsinghua professor who had been selected to serve as the university's deputy president, to join the party. Jiang then helped draw up a plan that identified 5,300 high intellectuals in Beijing and set a goal of recruiting 26 percent of them into the party by the end of 1957.[12] The plan was interrupted by the Anti-Rightist campaign in 1957, but it was resumed in 1958. Jiang, who had reasserted party authority in a particularly heavy-handed fashion at Tsinghua during the Anti-Rightist campaign, aggressively offered party membership to faculty members who were willing to criticize their dissident colleagues. Party leaders divided the Tsinghua faculty into "progressive," "middle," and "backward" groups based on an assessment of their political inclinations, and strongly encouraged progressive professors to apply for party membership.

Jiang's long-term goal was to turn the party organization at Tsinghua into a "professors' party." Before 1949, he said, the underground party organization had been composed of students, and since then it had evolved into an organization of young assistant teachers. For this reason, the party's authority at the university was still limited, and there was tension between the party and senior faculty. Tsinghua's full potential would only be realized once the university organization had become a professors' party. To accomplish this, Jiang declared, "Party members must become professors and professors must become party members" (*dangyuan jiaoshouhua, jiaoshou dangyuanhua*).[13] The party's goals, he said, could only be achieved by following a long-term strategy that involved "two types of people joining forces" (*liang zhong ren huishi*), using a martial term that invoked two military units converging to form a stronger force. The first type possessed political vision and power and the second possessed knowledge, he argued, and the second type was indispensable. "Our theoretical level is not high, our understanding is not deep enough, and we haven't accomplished enough," he warned party members. "We can't plan to just throw out everything old and think that heaven is going to send us people. Culture is by nature inherited; the greater part of learning and science are handed down by inheritance. We must admit that those who were educated under capitalist society have a greater mastery of knowledge. These intellectuals, however, can be reformed and serve socialism."[14]

The figures presented in Table 3.5 show that by 1965 Jiang had made significant progress in converting the university party organization into a professors' party. In 1948, there was not a single party member among the faculty. By 1951—the year before Jiang arrived at the university—forty faculty members had joined, including two of 169 full professors. By 1965, some 1,250 faculty members had joined the party, including seventy full or associate professors.

TABLE 3.5
Party membership among Tsinghua faculty by professional rank, 1948–1993

Year	Professors	Associate professors	Lecturers	Assistant teachers	Total
1948	0	0	0	0	0
1951	1.2%	7.5%	[13.3%]		8.5%
1965	[34.7%]		51.0%	53.0%	50.8%
1985	50.3%	58.5%	53.6%	53.8%	54.1%
1993	77.0%	67.6%	51.6%	31.6%	60.1%

SOURCE: Fang and Zhang (2001, vol. 1, 817–19).

By that time, over half the teaching staff were party members, including more than one-third of the senior faculty. The rate of membership was much higher among the faculty than among campus workers, of whom only 15 percent had joined the organization.[15]

Despite his vision of elite convergence, Jiang himself continued to distrust most of the old elite. He was especially skeptical about professors who had been trained in the United States, a segment of the faculty that had dominated the university before 1949. After Qian Weichang and other critics of the party were removed from positions of power during the Anti-Rightist campaign, the top university leadership did not include a single American-trained professor.[16] Even those members of the senior faculty who joined the party and who remained in important administrative positions never became part of the core party leadership, which included Jiang and six or seven deputy party secretaries. The members of this group were all veteran Communist cadres or former students who had led the underground Communist movement at the university before 1949.

When Communist troops arrived at Tsinghua in December 1948, the rapidly growing underground party organization had 205 student members, some of whom remained at the university and became the first generation of Tsinghua-brand cadres.[17] They were followed by students from subsequent cohorts who accepted Jiang's challenge to become "double-load" (*shuang jiantiao*) cadres, that is, cadres capable of exercising both academic and political leadership. Jiang was personally involved in the selection and cultivation of student cadres, and he scheduled time regularly to meet with them. "President

Jiang handled this very carefully," one student told me. "He had two standards—this guy has got to be very good in academics and also very devoted to him."[18] After graduation, many were hired by the university and quickly moved up in the school's administrative hierarchy. Jiang and the Tsinghua party organization preferred to rely on this new generation of Red experts, which they had selected and trained, rather than on converts from among the "old bourgeois intellectuals."

Cultivating a New Generation of Red Experts

When Jiang addressed the new students arriving at Tsinghua in the fall of 1965, he told them that in the future China would be run by engineers. Since the university was the country's leading engineering school, he said, it was likely that some of the students in the audience would become national leaders, and he expected that one day a Tsinghua graduate would be named premier. Although Jiang was elaborating a technocratic vision of socialism, no one in the audience could have mistaken his vision for that of Qian Weichang and the other intellectuals who had championed the rule of experts in 1957. Jiang was a strong advocate of party control, and in his vision only *Red* experts were qualified to govern.

RED AND EXPERT CULTURE

The CCP's incessant emphasis on combining Redness and expertise gave rise to an expectation at Tsinghua that outstanding academic and political qualities should coincide in the same individual, if not by nature, then as the result of exertion. Those who were politically committed and virtuous were expected to be the same individuals who excelled academically, and vice versa. This expectation was expressed at Tsinghua and other schools in numerous ways, one of which was the prestigious award "three-good student" given to a small number of young people with outstanding achievements in study, politics, and personal fitness. It was also evident every year at graduation time, when the *Tsinghua Bulletin* announced the proportion of the students graduating with academic honors who were also members of the party or the Youth League. The proportion was usually higher than the average for all graduates, an accomplishment highlighted in the announcements. The same expectation was displayed in reverse when Jiang expressed his disappointment about the comparative political backwardness of students selected to participate in the elite "teach according to ability" program, or the below-par academic performance of some of the students chosen to be political counselors.

The expectation that Redness and expertise should coincide was essentially a prescription for converting the Communist Party into a party of experts. In pursuing this goal, however, Jiang was careful to retain many of the party's traditions, ideological convictions, and class preferences. In the early 1960s, when he concurrently served as Tsinghua's president and minister of higher education, Jiang continued to promote elements of the education policies adopted during the Great Leap Forward and took particular pride in features that had been pioneered at Tsinghua, including the combination of teaching, research, and production in campus factories. Tsinghua students proudly donned blue work uniforms to do regular stints of training and production work in campus factories, and they also headed out to nearby communes to help bring in the harvest. Jiang stressed the importance of political classes and he personally taught classes on Marxist philosophy. Class line policies were moderated, but they continued to be enforced, particularly in political recruitment and in assigning students to high-prestige military-related disciplines.

Within the regime of tight political control that Jiang imposed on Tsinghua, however, there was room for children from a variety of class backgrounds to excel. Jiang wanted to train the best and the brightest, and he did not intend to mechanically exclude students who had problematic family backgrounds. To succeed, they had to be outstanding in their academic work, of course, and they had to make extraordinary efforts to prove their commitment to the party, which in their case included the onerous demand that they "draw a line" between themselves and their families. This requirement was an insurmountable obstacle for some. Cai Jianshe, a student at Tsinghua's attached middle school whose father, a prominent intellectual, had been denounced during the Anti-Rightist movement in 1957, was never able to comply. "When the class director asked me to break with my family, I could not do it," he told me. "That's very hard for a Chinese child to do; it caused a lot of pain for people."[19] This failure made it impossible for Cai to join the Youth League, restricting his career prospects. Other students were so determined to join the league and the party, however, that they did what was necessary to compensate for problematic family backgrounds, breaking with their families and making heroic efforts in the field of political performance.

The university party organization encouraged these students to overcome the fetters of their family background and it placed great emphasis on the role of manual labor in this process. Wei Jialing, whose father had been a high-level Nationalist Party official, recounted how she changed her world-view after she arrived at Tsinghua in 1958. At first, Wei recalled, she was upset because despite top examination scores she had been assigned to a department with little prestige. "Why just because my family origin was bad," she thought, "wouldn't they let me get into the department I selected?" Such personal considerations were left behind, however, once Wei and her classmates were sent off to help build the giant Miyun Reservoir. "[I learned] not to look down on working people, people whose bodies are dirty—they are the ones who are transforming the world," she said. "I learned that city people like me from bad backgrounds, we actually didn't know anything about working." Wei's recollections of the experience working on the reservoir were punctuated by heroic slogans of the period that promoted collectivism, gender equality, and integration with the workers and peasants. "We ate together, lived together, and worked together with the peasants and workers," she told me. "In the midst of 200,000 people working there, all the sudden I felt, 'One person is so small in the world!' It made me feel like becoming part of the collective and transforming the world."[20] Wei became an enthusiastic political activist, eventually overcoming the handicaps presented by her family origin and joining the party.

Many of Wei's fellow students were genuinely inspired by Communist ideals and were proud to push earth-filled wheelbarrows alongside workers and peasants. In spite of pervasive rhetoric about embracing the life of a "common laborer," however, it is unlikely that many thought of themselves as anything approaching "common." Tsinghua students always knew they were part of a highly select group, and this continued to be true even during the radically egalitarian days of the Great Leap Forward. Consider, for instance, this passage from an essay by a Tsinghua student, Xiao Lai, published in 1958 as part of the campaign to promote the Red and expert road: "We are carrying out the greatest socialist revolution and socialist construction in history and this task requires not only the enthusiastic labor of millions of workers and peasants, but also outstanding political and economic leadership talent, as well as a large number of technical experts. Only working-class intellectuals, only Red experts who wholeheartedly serve the working class and place their personal interests together with the interests of the great masses—not above the masses' interests—can accomplish this historic undertaking."[21]

Xiao Lai and other members of his generation of university students gladly accepted the appellation "working-class intellectual" and they embraced the idea of serving the people. Many willingly endured hardships, lived, ate, and worked together with the workers and peasants, and cultivated a pride in their ability to do heavy physical labor—all traits that certainly set them apart from traditional Chinese intellectuals. Nevertheless, it was possible to do all this and still have a profoundly meritocratic conception of one's special role in the world. This is apparent in the words of Xiao Lai, who we can imagine did not doubt that his future would be among the outstanding political and economic leadership talent required for the momentous tasks ahead. By enthusiastically joining his classmates as they helped villagers bring in the autumn harvest, a student could demonstrate his commitment to the principle of combining mental and manual labor, while at the same time preparing to serve society in a way more appropriate to his talents.

Despite this kind of ideological suppleness, some students noticed an inconsistency between the proletarian ideals that were the staple of their political education classes, in which workers and peasants were the masters of the country and revolutionary heroes had proletarian characteristics, and the technocratic and meritocratic ideals that animated campus life. Zheng Heping, a Tsinghua middle school student born into a family that included many highly accomplished intellectuals, attributed the school's meritocratic environment to the leadership of the principal, Wan Bangru, a highly regarded protégé of Jiang Nanxiang. "Back in the 1960s, the Communist Party wanted to treat everyone as a peasant, as a worker," Zheng told me. "But Wan Bangru . . . wanted everyone to become high class, like a professor or in a high field of engineering. . . . So that gave everyone, I think, a very deep influence. From very young, we thought of ourselves as a different people. . . . In Tsinghua's attached middle school we tried to let every kid become a genius—not average. You must be the people who contribute to the society—that is, the educated people. You must have something to feed back to society after you graduate, not just like normal people, working nine-to-five just to feed your family, that's all."[22]

COMBINING TWO ELITE WORLDS

Zheng and his classmates—or at least those who would succeed in becoming both Red and expert—would, indeed, become a different people, with characteristics that would distinguish them not only from the masses, but also from their own parents. Some of their parents had academic credentials and others had political credentials, but very, very few had both. This new generation would be able to claim both, and in the process of earning these credentials they would come to share a substantial body of common experiences and knowledge, and they would cultivate similar worldviews, values, cultural dispositions, and self-conceptions.

On the one hand, by attending elite schools, they were inducted into the rarefied world of the educated elite. This was a world of cultural refinement, which involved not only certifiable technical and literary expertise, but also worldliness, intellectual sophistication, and specific cultural preferences and manners of speaking, dress, and behavior (that is, the less formal aspects of cultural capital that so intrigued Bourdieu). This world was familiar to Tsinghua students who came from educated families, but it was much less familiar to those students whose parents were workers, peasants, and poorly educated revolutionary cadres. These students were joining a world that would remain foreign to their parents, as it would to the great majority of the population.

On the other hand, by joining the Youth League, they were inducted into the ardent political culture of the revolutionary Communist movement, which was now being converted into a culture of state cadres. This was a world of intense ideological commitment, collective endeavor, strict discipline, and public service. All Chinese citizens became acquainted to some degree with Communist political discourse (expressed in songs, stories, and texts) and ethical expectations, but members of the Youth League—and especially those who went on to join the party—lived and breathed this culture. This fervent world of Communist activism was familiar to those Tsinghua students whose parents were revolutionary cadres or soldiers, but it was foreign to those students who came from intellectual, working-class, and peasant families.

By 1966, over the course of nearly two decades of Communist power, the intersection of these two elite worlds had grown considerably. In the first two chapters, I described what happened when these worlds collided after 1949, as rough-hewn peasant revolutionaries moved from the villages to the cities and ensconced themselves uncomfortably alongside urban capitalists, professionals, specialists, managers, and officials in the upper echelons of Chinese urban society. The differences between the two groups decreased somewhat as peasant cadres took crash training courses (such as those offered at Tsinghua) and educated elites joined the party. Despite their new diplomas, however, the peasant cadres remained out of place in the refined world of the urban educated classes, and the relatively small number of urban elites who succeeded, through great diligence, in joining the CCP found that their new party cards could never make them revolutionaries or completely dispel the White aura they had acquired through their service to the old regime. Among the generations that came of age before 1949, the only individuals who were genuinely at home in both the Communist and the educated elite worlds were the relative handful of intellectuals, such as Jiang Nanxiang, who had joined the Communist insurgency.

Unlike most of their seniors, the Red and expert youth who were trained after 1949 were fully socialized in both worlds. They comfortably occupied the space at the intersection of these worlds, which the CCP—after its center of gravity had shifted from remote villages to urban office buildings—had provided with powerful institutional foundations. They had all excelled in school and had also fully embraced Communist ideology, and their experiences had given them a common sense of pride and purpose. They shared not only the perspectives and values imbued by the education system and the league and party organizations, but also a set of valuable credentials that gave them common interests and set them apart from the rest of society.

The solidarity of the Red and expert youth should not, however, be overstated. Their family origins were quite different—some came from the old educated elite, some from the new political elite, and others from more humble households—and individuals still bore traits that could be traced to their distinct backgrounds. These continued to provide potential lines of fissure, which broke open in moments of political turmoil.

Obstacles on the Red and Expert Road

All Communist leaders, including Mao Zedong, had reason to embrace the Red and expert logic. In order to consolidate Communist power and reduce their dependence on nonparty experts, it was necessary to train a new generation of experts who were committed to the Communist project and loyal to the party. It was easy to extend the Red and expert logic to make it the foundation of a grand technocratic vision—that the New China should be run by Red experts. The Communist commitment to rapid industrialization gave compelling force to this vision, and it found fertile soil in places like Tsinghua University, which was assigned to produce Red engineers. All Communists, however, did not share this technocratic vision. In fact, the prevailing political culture of the party presented intractable obstacles to realizing such a vision.

The Red-over-expert structure, created after 1949 by the sudden introduction of Communist cadres into top leadership positions in existing institutions, evolved into a permanent system. At Tsinghua and other schools, just as in government offices and economic enterprises, there were two distinct career tracks—one political/administrative and the other technical. The top leaders in every institution continued to be veterans of the Communist insurrection while other positions that entailed political and administrative power were entrusted to newly recruited party members. Those who had educational qualifications but had not joined the party could pursue a technical career, but party membership was required to advance up the political and administrative hierarchies. The distinction between the technical and political/administrative tracks was particularly pronounced in factories and other productive enterprises, where Communist leaders enlisted workers' help in running the enterprises and supervising the experts. Post-1949 university graduates were incorporated into this industrial structure in the subordinate position reserved for experts. These new experts were considered more dependable than the experts inherited from the old regime, but the new resembled the old in many respects. Most were from old elite families and even those who were not had taken on characteristics of the educated elite, including a tendency to prioritize technical over political qualifications and considerations. Even if a university graduate was a party member, therefore, he or she was considered a technician, not a potential leader. University graduates simply did not fit the prevailing image of a party leader, who ideally had emerged from the masses, still bore their rough-hewn features and spoke their language, could easily mingle on the shop floor, and was adept at attending to workers' problems, attenuating their grievances, and mobilizing them to take part in political and production campaigns. (Technical middle school graduates, who generally had more proletarian roots and attitudes than university graduates, were more likely to be considered leadership material.)

The obstacles that stood in the way of establishing a technocratic order can be glimpsed by considering the situation confronting Tsinghua graduates who were assigned to work in industrial enterprises in the 1950s and 1960s. Factory leaders—typically veteran cadres whose résumés included exemplary service during the revolution, but little education—gladly welcomed university graduates assigned to work in their enterprises, and after the new hires had completed a year or so of compulsory manual labor, they were eager to place them in technical positions in which they could put their valuable specialized training to work. Capable performance could lead to rapid promotion up a hierarchy of technical positions in the enterprise, which was headed by the factory's general engineer, a position of considerable responsibility and status. At each level (workshop/department, branch factory, factory, enterprise), however, ultimate power lay in the hands of the party secretary and the director. Capable and ambitious production workers could advance from group leader to shift leader to workshop director, or up the union and party hierarchies, and a few even made it to the upper echelons of the factory party organization and enterprise management, especially during the 1950s when political transition and rapid industrial expansion opened up many positions at the top.[23] Because the technical track was separated from the political/administrative track, however, university graduates could not expect to reach positions of power. Statistical data show the stark contrasts created by this Red-over-expert structure. On the one hand, factory leaders had far less education than the technical personnel who worked under them. A 1955 study reported that less than 6 percent of leadership personnel (*lingdao renyuan*) in China's industrial enterprises had a college education, compared to over 56 percent of general engineers and technicians (*zong gongchengshi, jishi*).[24] On the other hand, most managers were party members, while only a small minority of technical personnel joined the party. Even as late as 1965, according to a study of party membership in eleven large industrial enterprises, only 9 percent of technical personnel were party members, compared with 56 percent of management personnel. In fact, technical cadres were the least likely of all employees to be recruited into the party; over 18 percent of production workers had joined the party, double the membership rate among technical employees. As a result, technical cadres made up only 3 percent of the party membership in these enterprises.[25]

The party organization in China's industrial workplaces, typically led by revolutionary cadres of peasant origin and largely composed of members recruited from the ranks of the workers, presented a formidable obstacle blocking the technocratic path. The technocratic vision violated the egalitarian premises of Communist doctrine, which had been translated into a visceral antipathy toward the old elite classes and their pretenses to power based on cultural qualifications. This vision not only contradicted the political culture of the party organization, but it was also at odds with the practical interests of the bulk of party cadres and rank-and-file party members.

The Red-over-expert structure, which embodied the prevailing ideology and interests of the party organization, was militantly reconfirmed and reinforced during the Anti-Rightist movement and the Great Leap Forward, when experts who had challenged the power and prerogatives of the party—in industrial enterprises and government offices, as well as in schools—were harshly rebuffed. The party's position was expressed sharply in an article from this period published in the academic journal *Study*, which criticized arguments that "politics cannot lead technology," "non-experts cannot lead experts," "from now on technology will decide everything in the production struggle," and "technical intellectuals must educate the party." These arguments, the author wrote, reflected the reactionary ideology of technocracy, with the term "technocracy" rendered in Chinese as *gongchengshi zhuanzheng*, or "dictatorship of engineers."[26]

The most determined opponent of the technocratic vision was the party's top leader. Mao strongly supported the Red-over-expert structure and he shared Communist cadres' distrust of members of the old elite classes and their children. But he also increasingly distrusted the Communist cadres themselves, who were—in his estimation—too easily inclined to act like bureaucrats and aspire to the soft and privileged lifestyle of the bourgeoisie. His concerns about both groups—China's old and new elites—were already evident in 1957, when he first invited intellectuals to criticize bureaucratic tendencies among Communist officials, and then led the party's counterattack. In 1966, Mao would launch a new attack, this time aimed simultaneously at both groups. At the center of his crosshairs would be the Red experts, who had been so assiduously cultivated—with Mao's support—by the party organizations at Tsinghua and other elite schools.

PART TWO

The Cultural Revolution (1966–1968)

Chapter Four

Political Versus Cultural Power

On June 8, 1966, a work team composed of several hundred party officials arrived at Tsinghua with orders to suspend all university and department-level cadres and take charge of the school. The university had been in turmoil since the end of May, when a small group of radical teachers at Peking University had posted a caustic big character poster denouncing the school's leadership for practicing a "revisionist education line."[1] Mao Zedong endorsed the poster, and Tsinghua students flocked to the Peking University campus, which was within walking distance, to witness the ensuing controversy. Soon the Tsinghua campus was embroiled in a debate about the school's own leadership. Classes stopped and the walls of campus buildings were covered with big character posters attacking and defending the university administration. The work team, dispatched by central party leaders, authoritatively settled the debate by condemning Tsinghua Party Secretary Jiang Nanxiang and the entire university party committee, and it mobilized students and teachers to write big character posters and participate in meetings to criticize university leaders. Similar work teams were sent to other schools.

All of this was startling enough for members of the Tsinghua community, where Jiang's authority had been beyond question since he arrived at the university fourteen years earlier, but it hardly prepared them for what was to come. Within two months, Mao ordered the withdrawal of the work teams from schools and encouraged students, teachers, and staff to form their own "fighting groups" (*zhandou dui*) to bring down powerful party officials who, Mao said, were leading the country astray. Since the Tsinghua party organization had already ceased to function, the withdrawal of the work team left a power vacuum, which was soon filled by contending fighting groups, each with its own interpretation of the aims of the movement. This new political campaign, which Mao had dubbed the Great Proletarian Cultural Revolution, evolved into something quite unlike any of the other campaigns that had marked the first seventeen years of Communist power. Mao encouraged workers and peasants to join the movement, and by the end of the year the party organization around the country was paralyzed. It would remain largely inoperative for most of the next two years, and contending local coalitions, cobbled together out of small fighting groups, became the leading protagonists of the movement.

When Mao launched the Cultural Revolution, he directed fire at both of the groups that were uneasily sharing the top echelons of Chinese society—the old educated elite and the new political elite. The school system and the party organization, the institutional foundations from which these two elites derived their power, were both shut down and subjected to withering attacks. On the one hand, the social prestige, material position, and distinctive self-conception of intellectuals were attacked, often in ways that involved great cruelty to individuals. The practices of the entire education system were called into question, the categories of knowledge associated with traditional Chinese and Western culture—the patrimony of the educated classes—were denigrated, and symbols of these cultural traditions were ritually destroyed. At the same time, Mao called on students, workers, and peasants to attack party officials in their schools, workplaces, and villages. The authority of party offices was severely undermined as local officials were hauled up on stages to be criticized and humiliated by their subordinates.

Although Mao set the ideological and political premises of the Cultural Revolution, his pronouncements were often ambiguous and their meaning sharply contested. After the local party organizations ceased to function in the summer and fall of 1966, people in schools, workplaces, and villages across China coalesced into contending factions that pursued distinct and contradictory interpretations of the goals of the Cultural Revolution. Because Mao had targeted both political and cultural capital, each of these became a key axis of contention, with some factions attacking and some defending one or the other. This was clearly the case at Tsinghua, but factions were aligned in very different ways at the university and at its attached middle school. At the middle school, children of party officials battled children of intellectuals, in a conflict that pit political against cultural capital. At the university, in contrast, a "radical" faction attacked both the political and cultural pillars of the emerging class order, spurring the development of a "moderate" coalition—including children of both the old educated and the new political elites—that defended the status quo. Thus, the factions at the middle school represented a continuation of the long-standing conflict between China's political and cultural elites, while the moderate coalition at the university reflected nascent inter-elite unity. Both were manifestations of the contentious and protracted process of elite convergence.

Mao Zedong's Goals

Before examining how events developed at Tsinghua, it is necessary to consider the goals of the movement's paramount leader. In particular, Mao's decision to call on the masses to attack the officials of his own party requires explanation. Some scholars have explained this extraordinary turn of events as a product of Mao's efforts to enhance his personal power, while others—detractors as well as admirers—have described Mao as a utopian visionary determined to pursue Communist ideals of a classless society.[2] There is, of course, no necessary contradiction between explanations that emphasize personal power and those that stress ideology. Mao's vision of a Communist future was inseparably linked with a conception of his personal role in leading the people toward that future. Many scholars have argued convincingly that the Cultural Revolution reflected tensions between Mao's personal authority and that of the party bureaucracy.[3] Within the central party leadership, there was a division of labor, in which others handled day-to-day administrative affairs, while Mao was in charge of pushing forward the party's long-term goals, and he assumed responsibility for initiating major political movements.[4] These movements—including Land Reform, collectivization, and the Great Leap Forward—were instruments of revolution from above, which suddenly invoked transcendent Communist goals. Brief periods of calm were broken when Mao initiated new campaigns that violently overturned elements of the status quo, abrogating existing policies and institutions and creating new ones. Like the war chief in a tribe in which power is divided between a war chief and a peace chief, Mao's power was ascendant during moments of mass mobilization. During the Cultural Revolution, his personal charismatic authority reached its apex, based on appeals to realize the Communist mission of eliminating class differences.[5]

The official rationale for the Cultural Revolution was expounded in the thesis that the Soviet Union, China's model, was undergoing a process of "peaceful evolution" from socialism into a form of "state capitalism." According to Mao and a group of radical theoreticians associated with him, this transformation did not involve the fall of the Soviet Communist Party from power or a change in the ownership system. Instead, a new exploiting class was emerging within the leadership of the party, and its power was based on control over state and collective property. This theory had troubling implications for the Communist project. Prevailing international Communist discourse had emphasized the differences between socialism and capitalism: socialization of the means of production had eliminated exploitation and antagonistic class relations. The idea of peaceful evolution called into question this sanguine assumption. Instead of emphasizing differences, radical Chinese theorists began to emphasize the commonalities between socialism and capitalism. Despite the elimination of private property, the socialist system, like capitalism, was based on commodity exchange and wage labor, and distributed material goods unequally. Now, in their view, the Soviet officialdom had become an exploiting class—without any fundamental change in the social structure. Since China had closely followed the Soviet model, the Chinese social structure was also seen as harboring the seeds of exploitation, and the main danger to the Communist project came not from the overthrown propertied classes or from external enemies, but rather from "new bourgeois elements" *inside* the party.

To avoid following the Soviet path, Mao declared, it was necessary to carry out a "continuing revolution under the dictatorship of the proletariat." This revolution was to be directed against "those in power in the party who are following the capitalist road," condensed to the shorthand term, "capitalist roaders." In December 1965, on the eve of the Cultural Revolution, Mao warned that party officials were becoming an incipient exploiting class. "The class of bureaucrats is a class sharply opposed to the working class and poor and lower-middle peasants," he said. "These leaders who take the capitalist road have become, or are becoming, the capitalists who suck the workers' blood. How can they sufficiently understand the necessity of socialist revolution? They are the targets of our struggle and targets of the revolution."[6]

Mao's concern about a bureaucratic class emerging within the Communist Party did not, however, diminish his concern about the dangers posed by the old elites, and he was particularly worried about collaboration between the new and old elites. When he launched the Cultural Revolution in 1966, he identified both groups as the movement's targets. This was made explicit in the sixteen-point decision outlining the goals and methods of the movement, which was adopted by the party central committee—at Mao's insistence—in August 1966. The first point stipulated two principal objectives, "To struggle against and overthrow those persons in authority who are taking the capitalist road," and "To criticize and repudiate the reactionary bourgeois academic 'authorities.'"[7]

Going Around the Party

In previous political campaigns Mao had relied on the party organization; orders were passed down through the party's chain of command from the center to local branches, which mobilized subsidiary mass organizations, activating hundreds of millions of people. This time, however, because Mao's target was the party organization itself, he could not rely on it to lead the movement. Instead, he went around the party and directly mobilized students, workers, and peasants. The watershed event, which signaled the distinctive character of the new movement, occurred when Mao recalled the work teams that had initially been dispatched by party authorities to lead the movement.

Party work teams had long been used to lead political movements and rectify problems in local party organizations. During Land Reform, for instance, work teams spent months supervising the campaign in villages, making sure local Communist cadres were not protecting landlords and rich peasants. Work teams were also charged with investigating cadre corruption and abuse of power during the Socialist Education movement (1963–66). In these campaigns, work teams temporarily took charge of villages, factories, and schools—setting aside local party committees—and organized peasants, workers, and students to help investigate and criticize local party leaders. Work teams inspired great fear among local cadres, and the method was effective in enforcing party discipline and rooting out cadre corruption and abuse of power.[8] In 1966, therefore, it was quite natural for party leaders to assume that work teams would be the appropriate method to carry out Mao's latest initiative. This time, however, Mao was not simply seeking to discipline errant officials; he wanted to challenge the authority of the party organization. This was a task for which the work team method was ill-suited because it reinforced the underlying authority of the party hierarchy. Power was temporarily transferred to the work team, which represented higher party authorities, and then transferred back to (old or new) local party leaders after the work team left. During the entire process, the populace was expected to follow the direction of one set of party officials or another. The problem with the Socialist Education movement and previous efforts to reform the party, Mao concluded, was that they had all been directed by the party organization. In December 1965, in the midst of the movement, Mao declared that because members of the bureaucratic class were the targets of the struggle, "the Socialist Education movement can never depend on them."[9] After he had launched the Cultural Revolution, Mao noted: "In the past we waged struggles in rural areas, in factories, in the cultural field, and we carried out the Socialist Education movement. But all this failed to solve the problem because we did not find a form, a method, to arouse the broad masses to expose our dark aspect openly, in an all round way and from below."[10]

During the early months of the Cultural Revolution, Mao allowed party officials to send work teams to schools and workplaces, but he immediately undermined the authority of these teams by commissioning a series of newspaper and radio commentaries that condemned their efforts to control the movement and declared that "the masses must educate themselves and liberate themselves." This message incited confrontations in schools between students and work teams (and similar confrontations in factories), which eventually led to the emergence of a "rebel" movement that pledged loyalty to no one but Mao.

To lead the movement, Mao created a Central Cultural Revolution Small Group (CCRSG). CCRSG members typically shared two characteristics—ideological commitment to Mao's radical program and lack of power in the party organization. The group was led by Mao's wife, Jiang Qing, and his personal secretary, Chen Boda, and most of the other members were writers who had demonstrated a strong devotion to Mao's class-leveling agenda.[11] Although the CCRSG was formally an ad hoc committee attached to the party Central Committee, it answered to no one but Mao, and had no formal authority over any part of the party organization. Instead, it stood outside the party bureaucracy and led the attack against it.

No formal organizational links existed between the CCRSG and the myriad of local mass organizations (*qunzhong zuzhi*) that emerged at Tsinghua and elsewhere. These groups did not arise spontaneously; on the contrary, they arose in response to Mao's call. Nevertheless, the movement was not organized from above. Local groups were self-organized; leaders nominated themselves and gathered their own followings. The rise to prominence of specific local leaders and groups was sometimes the result of intervention by Mao and his lieutenants, and so mass organizations energetically appealed to them for recognition. No formal hierarchy of command, however, was ever established. The local organizations were themselves structured like political coalitions, reflecting their ad hoc origin. Small fighting groups coalesced into contending factions and they remained the basic units of the larger organizations. Membership in the fighting groups fluctuated as individuals joined and quit, and entire groups sometimes left one coalition to join another. Political activity was largely the work of these small, fluid groups. Their members discussed the issues of the day, collectively wrote big character posters, traveled across the country together, and—when factional contention turned violent—each group often procured or made its own weapons.

Because leadership of the movement was so loose, there was great variety in the factions that emerged in schools, workplaces, and villages across the country. Nevertheless, consistent patterns emerged. Although the factional alignments that developed at Tsinghua University and its attached middle school were very different from each other, the former had much in common with other universities, and the latter had much in common with other elite urban middle schools. In this chapter, I will analyze the political orientations, leadership, and social bases of the rival student factions at Tsinghua's attached middle school, leaving the factional alignment at Tsinghua University for the next chapter.

Birth of the Middle School Red Guards

On the evening of May 29, 1966, a dozen students from Tsinghua's attached middle school—most of whom were children of Communist cadres—met secretly in the vast, overgrown gardens of the Old Summer Palace, located next to the school. At the meeting, they decided to create their own organization, which they called the Red Guards, and took an oath to fight to the death to defend Mao Zedong Thought. Although most of them were members of the school's Youth League, they thought it was not militant enough. The league, they believed, was subservient to teachers and school leaders, who were too concerned about academics and not enough about class struggle. In particular, they thought, teachers and school leaders were overly accommodating to students who were academically strong but came from old elite families, allowing them to join the league and even allowing them to be elected to leadership positions, in violation of class line principles. The small group discussed the exciting events at Peking University, and they decided the time was ripe for them to denounce their own principal, Wan Bangru.[12]

Thus was born China's first Red Guard organization. The name would eventually become a generic designation for student organizations during the Cultural Revolution, but the first wave of Red Guards was a distinct political tendency with a peculiar agenda that set it apart from subsequent student factions. The group at Tsinghua's attached middle school not only invented the name, but also served as a model that was soon replicated by audacious teenage children of Communist cadres in other Beijing middle schools.

At Tsinghua's attached middle school, the fledgling Red Guards posted big character posters lambasting Principal Wan, denouncing his landlord family origin, and accusing him of implementing revisionist education policies. Soon after these provocative posters appeared on school walls, other students—including many from intellectual families—replied with posters defending the principal and the school administration. The Red Guards responded with new posters calling their opponents "royalists" (*baohuang pai*) for coming to the aid of the principal. The insurgents, however, were in the minority and—under attack by the administration and its student defenders—they retreated from the school on June 7. They returned the next day, bringing with them a bicycle-mounted contingent of scores of students from other elite Beijing middle schools, most of whom were also children of revolutionary cadres. School officials locked the gate and a shouting match ensued between students defending the principal, standing inside the school, and the Red Guards on the outside. That evening, a work team dispatched by Beijing party authorities arrived and deposed the school's leadership. On the heels of the work team, the Red Guards reentered the school triumphantly.[13]

The battle lines that would define the student factional conflict at the middle school for the next two years were drawn in this first confrontation. The school split along family origin lines, with students from revolutionary cadre families (who made up about one-quarter of the student body) standing on one side, and students from intellectual families (who made up about two-thirds of the student body) standing on the other. Before the Cultural Revolution, Youth League activities had involved students from both groups, and friendships had crossed family origin lines. Tensions between the two groups, however, had increased over the previous two years, as Mao called on young people to remember class struggle, and by the summer of 1966, the school polarized into two antagonistic camps.[14]

During the remainder of June and July, the work team was in charge of the middle school. Classes were canceled and students were mobilized to criticize the administrators and teachers by writing big character posters, engaging in small-group discussions, and participating in mass "criticism and struggle" meetings. In selecting students to help lead the movement, the work team followed class line principles. Students from intellectual families who had been Youth League leaders were now not only left on the sidelines, but were attacked for having supported the old administration. The work team asked the Red Guards, now heroes for their early broadsides against the principal, to preside over schoolwide meetings.

When Tsinghua University students demanded that the work team leave campus and let students organize themselves (see Chapter 5), middle school Red Guard leaders did not join the fray. It was one thing to attack a middle school principal who—despite being the school's party secretary—was from a landlord family and could easily be identified as a "bourgeois academic authority." It was quite another to attack work team members who had been dispatched by the highest party authorities and who personally had impeccable revolutionary credentials. When one member of the middle school Red Guards posted a big character poster defending the insubordinate university students, her comrades abandoned her to the retribution of middle school work team leaders, who branded her a "counterrevolutionary."[15]

ATTACKING CULTURAL CAPITAL AND DEFENDING POLITICAL CAPITAL

At the end of July, Mao ordered the removal of work teams from schools, and then on August 1, he sent an open letter to the Red Guards at Tsinghua's attached middle school praising their rebellious spirit.[16] With Mao's endorsement, which was reaffirmed when he greeted millions of students who gathered for rallies in Tiananmen Square, the Red Guard movement swept the country. Students in schools across China, typically with revolutionary cadres' children at the head, formed their own Red Guard organizations.

With the school party and Youth League organizations paralyzed and the work team gone, the young Red Guards at Tsinghua's attached middle school eagerly took over responsibility for leading the movement. Their mission, as they understood it, was to attack the old cultural elite and the education system. Liao Pingping, the daughter of a peasant revolutionary who had become a high-level official, explained the organization's aims as egalitarian. "The Red Guards were against inequality; we believed we were following the past ideals of the Communist Party—helping the poor," she told me. "We demanded they open the school doors to worker and peasant children." In Liao's view, she and her friends were fighting against elitist education. "We criticized the school leadership and the education system; we thought Tsinghua's integrated system was wrong—studying straight through from Tsinghua's attached primary school directly to Tsinghua University—that was not right; there was no connection with society. . . . We criticized the school for putting too much pressure on students. In grades, in sports—everything was about competing to be the very best."[17]

All of these points were elements of Mao's critique of the education system, which he had been pressing with renewed vigor since his 1964 Spring Festival speech. Many revolutionary cadres' children had made Mao's radical education agenda their own long before the Cultural Revolution began. In 1965, the students at Tsinghua's attached middle school who would later create the Red Guards organized an "Education Reform Group" and they wrote a series of big character posters criticizing the school for not following Mao's education line. In 1964, a similar group of students at Beijing's elite Number 4 Middle School went on strike to demand education reform, and they later wrote a letter to the party Central Committee calling for the elimination of college entrance examinations and implementation of stricter class line admissions policies.[18]

By July 1966, with the Cultural Revolution at full tilt, the Red Guards at Tsinghua's attached middle school railed against the education system and the old educated elite with passion and adolescent flair. "For seventeen years our school has been ruled by the bourgeois class. We shall not tolerate this any longer! We shall overthrow it, seize power, organize the revolution of the class troops, forward the class line according to social status." They directed their wrath particularly against their classmates from old elite families. "Landlord and bourgeois class young gentlemen and ladies, we know your feelings. . . . You thought you could make use of the temporarily existing bourgeois education to climb higher up the ladder to become White experts, get into the university, join up with 'professors, experts.' . . . And perhaps you could even build up a little political capital and get a little power. . . . Truly you did not imagine that the class line that you hate would come and destroy these dreams. . . . Workers and peasants and the children of workers, peasants and revolutionary cadres, whom you despised, will fill the posts in culture, science and technique. Your monopolies are broken."[19]

Although the Red Guards attacked the old educated elite in the name of the workers and peasants, their organization was actually composed almost exclusively of revolutionary cadres' children. There were, after all, very few children from working-class and peasant families at the middle school, and most of them were rejected by, or alienated by, the Red Guards.[20] Revolutionary cadres' children at elite Beijing middle schools developed their own peculiar interpretation of the party's class line, known as "bloodline theory" (*xuetong lun*), which highlighted their own role. The bloodline principle was expressed concisely in a famous couplet, created by students at another elite Beijing middle school, that became the Red Guards' motto: "The father's a hero, the son's a brave lad; the father's a reactionary, the son's a bastard" (*laozi yingxiong er haohan, laozi fandong er hundan*). Red Guards at Tsinghua's attached middle school were convinced—as they put it in a flier—that they had a unique responsibility "to follow in the footsteps of our revolutionary fathers."[21] Before Mao launched the Cultural Revolution, he made it clear that he hoped a new generation of "revolutionary successors" would be forged in the upcoming battles. "The cadres' kids thought they naturally were revolutionary successors," recalled Song Zhendong, a student at Tsinghua's attached middle school whose parents were both veteran revolutionaries.[22]

The ascendance of bloodline theory in the summer of 1966 politically marginalized the great majority of students at Tsinghua's attached middle school. Several students from intellectual families who had been close friends with Red Guard leaders had joined the organization in its early days, but the rise of bloodline theory compelled them to withdraw, as membership now required revolutionary origins. A Red Guard activist estimated that by August about 90 percent of the students from revolutionary cadre families had joined the group and they made up virtually its entire membership. The organization claimed three hundred members, about 20 percent of the student body. Shortly after the work team left, the Red Guards issued a flier declaring, "All those who are not children of workers, peasants or revolutionary cadres . . . had better lower their heads before us!"[23]

In August, Red Guard organizations at elite Beijing middle schools became the vanguard of a violent campaign against anyone who could be identified with the old elite classes, including administrators, teachers, and students in their own schools. At Tsinghua's attached middle school, several administrators and teachers were severely beaten; the principal suffered a damaged kidney, a teacher lost an eye, and another teacher committed suicide after enduring violence and humiliation at the hands of students.[24] At the same time, the Red Guards led the campaign against the "four olds" (old ideas, culture, customs, and habits), searching the houses of the old bourgeoisie and prominent intellectuals, interrogating and sometimes beating the inhabitants, and destroying or confiscating symbols of traditional elite culture or Western influence.

While the children of revolutionary cadres enthusiastically took up Mao's call to attack the old cultural elite, they were not as enthusiastic about his simultaneous call to attack the new political elite. In fact, for them, this aspect of Mao's program was incomprehensible. They interpreted Mao's agenda in a manner that made sense to them: attack the old elites in order to defend the Communist Party. This was the agenda they were following on August 24, 1966, a day of violence that marked the apex of their power. The day before, Red Guards at Tsinghua's attached middle school discovered that Tsinghua University students had put up big character posters criticizing Liu Shaoqi, the CCP's organization chief, who had been responsible for dispatching work teams to schools in June, and his wife, Wang Guangmei, who had been a prominent member of the work team at the university. They mobilized Red Guards from twelve middle schools around the city to converge on Tsinghua. After declaring at a mass rally that they would never allow reactionary students to "turn over the country" (*fan tian*), they used a truck to pull down the university's famous gate, a symbol of the school's prerevolutionary heritage. That afternoon and evening, they went on a rampage, beating up administrators and professors who had been criticized as bourgeois academic authorities or members of the "black gang" responsible for carrying out revisionist education policies. At the same time, they tore down the offensive wall posters and attacked university students who had criticized the work team and the central party officials responsible for dispatching it.

During the coming weeks, Red Guard activists were astonished and dismayed as they came to realize that Mao and his lieutenants in the CCRSG supported the student groups that were attacking the work teams and party officials. By October, the students were on the defensive, trying to stave off attacks against themselves and their parents. In December, Red Guards from elite Beijing middle schools convened a mass meeting to establish a Capital Red Guard United Action Committee. Hundreds of students from Tsinghua's attached middle school attended, along with thousands of other middle school students. The loose-knit alliance that came out of the meeting, called United Action for short, openly opposed the CCRSG, and a flier issued in its name pledged to defend "the party organizations at all levels and the outstanding, loyal leading cadres."[25] Liu Jinjun, one of the principal leaders of the Red Guards at Tsinghua's attached middle school, told me: "By then [we] were not fighting for Mao anymore, we were fighting for survival."[26]

A New Rebel Movement

By late fall 1966, the Red Guards at Tsinghua's attached middle school were being challenged by a new student organization, Jinggangshan, named after the mountain stronghold from which Mao and his confederates had launched their rural guerrilla strategy nearly four decades earlier. The organization was composed mostly of students from intellectual families, who made up the great majority of the school's student body. Before the arrival of the work team in June, many of the students now active in Jinggangshan had defended the school's principal, and the Red Guards—styling themselves as rebels—had denounced them as royalists. Now, the intellectuals' children claimed the rebel mantle for themselves and hurled the royalist charge back at their Red Guard adversaries, accusing them of defending the capitalist roaders in the party.

These new middle school rebels were allied with a much larger student group at Tsinghua University, from which they took their name. The Jinggangshan organization at the university was founded by students who had challenged the authority of the party work team, a move that had subjected them to harsh criticism at the time, but had made them heroes after the work teams were recalled (see Chapter 5). On October 6, 1966, Jinggangshan and like-minded student organizations at other Beijing universities organized a massive rally, reportedly attended by over 100,000 people, at which they denounced the suppression of students by the work teams in June and July. They demanded that work team leaders be held accountable, and extended their attack to include the higher-level officials who had dispatched the work teams. Several members of the CCRSG spoke at the rally, pronouncing their support for this new movement. The rally marked a turning point in the Cultural Revolution. Until then, under the leadership of the work teams and the Red Guards, the movement had been directed mainly against the education system and the cultural elite. This new wave of rebel organizations, with the support of the CCRSG, made the party organization and the political elite its main target. Cai Jianshe, a Jinggangshan activist at Tsinghua's attached middle school, summed up the change this way: "First Mao used the Red Guards to attack the reactionary academic authorities, and then he used us rebels to attack the capitalist roaders."[27]

Jinggangshan activists at Tsinghua's attached middle school claimed that at their strongest moment, in the winter of 1966–67, they had the support of 80 percent of the school's students. Although this estimate is probably exaggerated, the new group was by all accounts much larger than the Red Guards had ever been. Complying with the class line norms of the Cultural Revolution, the group's two main leaders were children of revolutionary cadres (the key leader was the former Red Guard who had been abandoned by her erstwhile comrades after she supported the students who criticized the work team at Tsinghua University). Nevertheless, the organization belonged to children of intellectuals, who made up virtually its entire membership.

ATTACKING POLITICAL CAPITAL AND DEFENDING CULTURAL CAPITAL

The debate between the two student factions at Tsinghua's attached middle school often involved national policies and ideological principles, but even at its most abstract moments the struggle was intimately linked with the personal status of members of the contending parties. The revolutionary cadres' children denounced their adversaries as scions of the old cultural elite, and the intellectuals' children denounced their adversaries as scions of the new political elite. As the two sides traded charges of privilege, Jinggangshan activists were as caustic and righteous as their opponents. In January 1967, they denounced United Action, the loose coalition that included the Red Guards from Tsinghua's attached middle school, as defenders of a new privileged stratum. "United Action is composed of a group of cadres' children whose outlook has not yet been reformed," they declared. "Because they are in an advantageous political and economic position, [their parents]—especially those revisionists—try to inculcate in them the idea that they deserve political and life-style privileges in order to train their children to be revisionist sprouts and later take over their positions. [These revisionists] allowed their children to be divorced from labor and from workers and peasants and inculcated in them the idea of being 'born Red' (*zilai hong*). They encouraged them to abandon thought reform and efforts to be sincere children of the people; instead they are climbing up above the people and becoming an intellectual aristocracy (*jingshen guizu*). These children are revisionist sprouts; they are the successors of the privileged stratum."[28]

The rebels embraced precisely that part of Mao's program that the Red Guards found incomprehensible: his call to attack the new political elite and combat bureaucratic privilege and power. At the same time, they had much less interest in Mao's call to radically transform the education system and attack the old cultural elite. Cai Jianshe, the Jinggangshan activist, made this point succinctly. "What we didn't like was privilege (*tequan*)," he told me. "We never opposed knowledge."[29] Several Jinggangshan leaders had, in fact, been among the most resolute defenders of the middle school principal in June. Six months later, these same students were compelled to accept the verdict that the school administration had been following a revisionist education line, but they were hardly ardent opponents of the old education system and those who ran it. "We supported changes in education," a middle school rebel leader told me, "but we were not so violently opposed to the principal and the school leadership."[30]

The most immediate and visceral issue for middle school rebels was the Red Guards' bloodline theory. This issue was a central focus of debates on middle school campuses from the fall of 1966 through the spring of 1967. Members of the CCRSG joined the debate, strongly condemning the idea of "natural Redness" claimed by revolutionary cadres' children, but defending class line preferences. In so doing, they were reflecting the position of Mao, who was more adamant than ever about preventing the reproduction of the educated elite. Many rebel activists at Tsinghua's attached middle school, however, believed that all class line policies were unjust. These students felt their own views were expressed in an essay titled "On Family Origin" penned by Yu Luoke, a young employee in a Beijing factory whose bourgeois family origin had prevented him from attending college despite excellent examination scores.[31] Yu's bitter manifesto, printed in a newspaper published by students at the elite Number 4 Middle School in January 1967, condemned the entire family origin system, which he likened to a caste system. Although those of "good" class origin claimed they were discriminated against, Yu wrote, they had actually received special treatment. It was those of "bad" class origin who faced discrimination, he added, providing a host of examples. "We do not recognize any right," he declared, "that cannot be attained through individual effort."[32] Yu's eloquent defense of meritocratic principles hit a responsive chord among students from educated families. They had long believed that school admissions should be based on examination scores and Youth League and party recruitment should be based on an individual's political performance, not his or her family origin. "On Family Origin" was widely debated in student factional newspapers and over one million copies were reprinted and distributed around the country during the winter and spring of 1967.[33]

LITTLE ROOM FOR VOICES OF MODERATION

In the winter of 1967, the polarized factional alignment at Tsinghua's attached middle school was complicated by the emergence of a third faction, the Mao Zedong Thought Red Guards, which stood for moderation. The students who founded this organization had been appalled by the Red Guards' attacks on the school principal and the teachers and they were also equally disturbed by Jinggangshan's attack on the political authorities. They thought the status quo before the Cultural Revolution was not that bad and they had no inclination to attack either the cultural or the political elites. While both Jinggangshan and Red Guard activists considered themselves rebels (and disparaged their opponents as royalists), the moderates were not interested in any kind of rebellion. They were, in the words of a former Jinggangshan activist, "fundamentally royalist."[34]

The moderate faction was composed mostly of children of intellectuals, but their ranks were joined by revolutionary cadres' children who abandoned the Red Guards after bloodline theory fell into disrepute. Among the moderate activists were several members of the schoolwide committee of the now defunct Youth League. Indeed, the faction resembled the Youth League in both its political stance and its membership: it supported the establishment—both political and academic—and it was composed of children of both the new political and old educated elites. The moderates, however, did not play an important role in the factional conflict at Tsinghua's attached middle school. Despised as insipid and spineless by combatants in the rival Red Guard and Jinggangshan camps alike, the moderates were a small and ineffectual group.

Despite the emergence of the moderate faction, the school remained largely polarized into two camps, with children of the new political elite battling children of the old educated elite.

Breaking Open Fissures Among Red and Expert Youth

The continuing fragility of the postrevolutionary accommodation between China's new political and old educated elites was clearly evident at Tsinghua's attached middle school in 1966. The school was populated by children from both of these groups. They had much more in common than their parents did, as they had all been immersed in both the worlds of scholarship and of Communist political activism. Almost all of the students who came from revolutionary cadre families—including those whose parents were not highly educated—had tested into Tsinghua's attached middle school, and some were highly successful in the academic competition. In fact, the founders of the Red Guards were members of the school's special college preparatory class, which required stellar academic performance. At the same time, most students who came from intellectual families had already joined or could look forward to joining the Youth League, and many had achieved leadership positions. Nevertheless, after Mao made targets of both the political and educated elites in 1966, the school quickly split into antagonistic factions based on family origin.

The young Red Guards became the primary agents of this split. Revolutionary cadres' children enthusiastically responded to Mao's call to attack the educated elites, and it was bloodline theory—their interpretation of Mao's class line policies—that caused middle school students to coalesce around polarized sets of interests and identities. On both sides, interests were reinforced by moral convictions. On one side, revolutionary cadres' children were aware that they enjoyed distinct advantages provided by intimate links to the ruling party, but they also saw themselves as defending a just social revolution, which had overturned a highly inequitable class order. On the other side, intellectuals' children knew that selection by examinations provided them with important advantages, but they were also moved by moral indignation toward bloodline theory, which represented the aristocratic tendencies of a new system of political privilege. Such privileges violated not only Communist principles of social equality, but deeper meritocratic conceptions of right and wrong long cherished by Chinese intellectuals. Each of the middle school factions ardently embraced one aspect of Mao's agenda, with revolutionary cadres' children attacking the academic establishment and intellectuals' children attacking the political establishment. Students split along family origin lines, and the ensuing conflict broke up fledgling connections between the two groups and hardened divisions between them. Within such a polarized environment, there was little room for a moderate faction that defended both the political and academic establishments.

The factional alignment at Tsinghua's attached middle school reproduced the battle lines of 1957, when the old educated elite and the new political elite were contending for power. The situation at the school was not at all unique. In fact, the most widely accepted analyses of student conflict during the Cultural Revolution have highlighted contention between children of revolutionary cadres and children of intellectuals.[35] The present analysis corroborates the main thesis of these studies—that Cultural Revolution battles reflected ongoing discord between China's new and old elites. These studies, however, were based largely on investigations of elite urban middle schools similar to Tsinghua's attached middle school. As we shall see in the next chapter, the nature of the conflict at Tsinghua University was very different from that at at its attached middle school, and shifting factional alignments at the university ultimately gave rise to inter-elite unity.

Chapter Five

Uniting to Defend Political and Cultural Power

During the summer of 1968, the Tsinghua University campus was divided into two sections, each occupied by a different student faction. Students on both sides were armed with spears and other weapons manufactured in campus machine shops, as well as a small number of firearms. That spring, political contention had escalated into violent confrontations as students fought for control of university buildings. After a series of skirmishes, buildings in the northern and western part of the campus were in the hands of one faction, and buildings in the southern and eastern part of the campus were in the hands of another. For most of the summer, the Science Building, held by one faction but located in the territory of the other, was under siege and, at one point, students surrounding the building tried to drive out those holed up inside by setting the building on fire. By the end of the summer, twelve students had been killed and dozens of others injured.[1]

Students in both camps were convinced they were fighting for socialism. On one side, the "radicals" were answering Mao Zedong's call to overthrow the capitalist roaders in the party in order to prevent the restoration of capitalism. On the other side, the "moderates" were defending the Communist Party and the existing socialist order from would-be usurpers.[2] Superficially, the daily newspapers and fliers published by the two groups looked very similar: both were emblazoned with red flags, photographs of Mao, and headlines defending Mao Zedong Thought and denouncing revisionism. The entire campus community, however, was acutely aware of the differences between them. The radicals were determined to "thoroughly smash the old Tsinghua," and they directed their fire at the university party organization and its policies. The moderates were defending the status quo and the university establishment.

The two factions at the university, unlike those at Tsinghua's attached middle school, could not be easily distinguished in terms of the family origin of their members. Students from intellectual, revolutionary cadre, worker, and peasant families could be found in both camps, and the principal leaders of both organizations were students of peasant or worker origin. Thus, the factional conflict at the university was not a struggle *between* classes, in the sense that one class lined up against another. Nevertheless, as I will show in this chapter, it was a struggle *about* class, because at stake were the political and cultural foundations of the emerging class order.

The Rise of the Rebel Movement

Soon after the work team arrived at Tsinghua in early June 1966, Kuai Dafu, a twenty-one-year-old chemical engineering student, wrote a series of big character posters accusing work team leaders of trying to control the student movement. The work team, which was headed by the deputy chairman of the State Economic Commission, Ye Lin, and counted among its members President Liu Shaoqi's wife, Wang Guangmei, had locked the university gates, prohibited contact between students from different departments, and required that big character posters be approved in advance. In his unapproved posters, Kuai called for the expulsion of the work team from campus.

Looking at his résumé, Kuai was an unlikely rebel. He had been very active in party-led student political activities at Tsinghua and his family origin was impeccable. His father, a peasant who joined the party before 1949, had served as bookkeeper and deputy secretary of the party branch in his village production brigade.[3] His mother had also been active in the underground Communist movement and joined the village party branch in 1954. Kuai excelled in school and in 1963 he was one of the few students in his rural county to test into college, with examination scores that allowed him to enter Tsinghua. He was invited to join the prestigious Department of Chemical Engineering, a discipline generally restricted to students of "good" class origin because it was related to missile technology. He soon became a leader of the Youth League branch in his class, a company leader in the school militia, and director of the editorial committee of the university broadcasting system. Inspired by the Socialist Education movement, in 1964 Kuai sent a report to the People's Congress exposing corruption among cadres in his home county, and he published an article in Tsinghua's Youth League newspaper criticizing his uncle, a commune cadre.[4]

Kuai had almost completed the protracted process of joining the party when he posted his criticisms of the work team, taking the fateful first step on a path that would make him into an implacable enemy of the party organization. "I didn't like the work team's methods," he told me. "The newspapers said it should be a students' movement, but the work team wanted to control everything very closely. That's not what Mao Zedong was urging us to do. . . . Liu Shaoqi . . . didn't understand Mao's thinking; he thought the universities were very chaotic, so he sent work teams to try to control the situation. The work team . . . suppressed the students."[5]

On June 24, the work team convened a campus-wide meeting to criticize Kuai, condemning him as a "counterrevolutionary." Two Tsinghua students, President Liu Shaoqi's daughter, Liu Tao, and Field Marshal He Long's son, He Pengfei, were selected to preside over the meeting. An unrepentant Kuai denounced the work team, winning loud applause from perhaps half of the thousands of people crowded into and around the school's main auditorium. Students and teachers, who were accustomed to the tightly controlled political environment at the university before the Cultural Revolution, were astonished by Kuai's defiance. "Then you couldn't doubt the leaders, so it became a big deal," explained Ke Ming, a student who decided to support Kuai. "That changed during the Cultural Revolution—then you could. That was the impact of Mao Zedong Thought. The extraordinary thing about Kuai Dafu was that he saw that back then, and he didn't back down." Zhang Youming, a student who had been selected by the work team to help control access to the stage, ended up siding with Kuai. "I didn't know who was wrong or right, but I felt the cadre kids and the work team didn't let Kuai Dafu express himself, so I stopped the cadre kids and the work team people [from approaching the stage] and I helped Kuai Dafu," Zhang told me. "I felt that if it was a debate, then both sides should have the freedom to speak." The meeting was a decisive moment; the campus split into two incipient factions, one supporting and one opposing the work team.

The work team mobilized students to criticize classmates who had supported Kuai, labeling them "Rightists" and "counterrevolutionaries." It was never able, however, to reimpose the kind of control that had existed before the movement began. Then in late July, Mao ordered the work teams removed from schools, and a few days later he issued what he called his first big character poster, titled "Bombard the Headquarters." In the poster, he sharply denounced the methods of the work teams: "In the last fifty days or so some leading comrades from the central down to the local levels have . . . [adopted] the reactionary stand of the bourgeoisie, they have enforced a bourgeois dictatorship and struck down the surging movement of the Great Proletarian Cultural Revolution. They have stood facts on their head, juggled black and white, encircled and suppressed revolutionaries, stifled opinions differing from their own, imposed a White terror, and felt very pleased with themselves."[6]

Before the work team left Tsinghua, it hastily appointed a Cultural Revolution preparatory committee to take charge of the movement, which was led by Liu Tao, He Pengfei, and other students whose parents were top party officials. These students soon formed a Red Guard organization, which established close links with the Red Guards at the university's attached middle school. He Pengfei had recently graduated from Tsinghua's attached middle school and he knew the leaders of the middle school Red Guards well. Like their middle school counterparts, the university Red Guards employed harsh class line rhetoric and concentrated their attacks on the old elite, the academic establishment, and symbols of traditional culture. On August 24, they joined middle school Red Guards in pulling down the famous university gate and attacking university professors and administrators.

Unlike their middle school counterparts, however, the Tsinghua University Red Guards had competition from the beginning. While the extreme class line requirements that prevailed during the early months of the Cultural Revolution restricted participation in Tsinghua's attached middle school to children of Communist cadres (as almost all of the other students were from intellectual families), nearly 40 percent of the students at the university were from working-class or peasant families. They had the class qualifications to participate, and while many followed the lead of the revolutionary cadres' children in the Red Guards, others formed their own fighting groups. The campus split into two factions, defined by their stand toward the work team—the Red Guards, led by the cadres' children, defended the work team and their opponents attacked it. The underlying question was whether or not the party organization should control the student movement.

Kuai Dafu and several of his classmates established their own small fighting group, Jinggangshan, which eventually became the leading group in the anti–work team camp. Kuai had received a visit from leaders of the CCRSG even while he was under investigation by the work team in July, and later the CCRSG encouraged Kuai and leaders of similar anti–work team groups at other universities to form a citywide coalition. This coalition, known as the Third Red Guard Headquarters, harshly criticized not only the work teams, but also the higher officials who had sent them, and all authorities who had suppressed independent activity by students and workers in recent months. Kuai and his confederates in the Third Headquarters condemned all such repression as a "bourgeois reactionary line" designed to protect party officials, and they charged the early Red Guard organizations with being accomplices in carrying out this repression. Kuai addressed the massive October 6 rally in Beijing that launched the offensive against the bourgeois reactionary line and gained national fame as a leader of the new rebel movement. Tsinghua students rushed to join his organization.

After the work team left, Tsinghua became the site of frenzied political activity. Impassioned debates took place in mass meetings and at impromptu gatherings, layers of big character posters covered campus walls, the public address system was commandeered to announce meetings and hurl invective at opponents, and fliers delivered the manifestos of new organizations. After Mao called on young people to travel around the country to "link-up" with others and "exchange revolutionary experiences," Tsinghua became a destination for tens of thousands of itinerant youth, who camped out at the university and participated in the debates. Tsinghua students also traveled around the country, joining millions of young people crowding China's railroads and highways. Mao insisted that local authorities welcome the travelers and provide them with transportation, food, and lodging without charge. These emissaries of rebellion not only went to other schools, but also to factories and villages, spurring the formation of local rebel groups and making certain that no local party committee would escape unchallenged.

In January 1967, Mao called on the fledgling rebel organizations to seize power (*duoquan*) from party authorities in schools and workplaces across China. The authority of the party organization—which before the Cultural Revolution could not be challenged—was broken, and rebel groups tenuously took charge of Tsinghua and many other schools and workplaces. The fate of individual cadres was debated at mass meetings, in which participants evaluated their self-criticisms and discussed who among them was fit be restored to leadership positions. This extraordinary turn of events was the result of the combined efforts of Mao at the top and the rebel organizations at the bottom. Mao and his rebel followers shared the goal of undermining the power of the officials who staffed the party organization in the middle, and they depended on each other. Without the rebels at the bottom, Mao's crusade against the party bureaucracy would have had little impact, and without Mao's support at the top, the rebels could not have survived. The dynamics of this top-and-bottom-versus-the-middle strategy were evident in Kuai Dafu's first big character poster denouncing the work team, in which he wrote: "We will oppose anyone who opposes Mao Zedong Thought, no matter how great his authority or who he is."[7] Kuai's manifesto was both an unprecedented challenge to the authority of the party hierarchy and an expression of unstinting loyalty to the supreme leader of the party (or, more precisely, to the charismatic mission expressed in his thought). Kuai used his loyalty to the supreme leader as a weapon to challenge the party officials above him.[8]

Except for a six-week period during the 1957 Party Rectification campaign, participation in previous political movements since 1949 had always entailed following the guidance of the party hierarchy. Participation in the Cultural Revolution, in contrast, entailed following Mao's personal leadership. Mao enjoyed tremendous power and he could change the course of events by uttering a few words. But the Great Helmsman was a distant leader and his words were few. Once the party organization had been paralyzed, people gained unprecedented power to think and act independently. Ke Ming, the student leader at Tsinghua, described how this type of mobilization undermined the authority of the party hierarchy. "Before the Cultural Revolution, everything came down from above, one level at a time," he explained. "You had to listen to those right above you. Then, all of the sudden, Mao went around the hierarchy and told the masses that the people between him and them had problems; that they should not listen to them. 'Think for yourselves.' This was the first time we had room to think for ourselves. That's why we supported the Cultural Revolution."[9]

The Mao personality cult reached its height during the Cultural Revolution. His image—associated with a red sun that conjured up divinity—was ever present and his words were imbued with infallibility. Although Mao expressed his discomfort with extreme manifestations of this "individual worship," there can be no doubt that it served to reinforce his personal authority at a time when he was using this authority to challenge the authority of the party organization.[10] The rebels were just as dependent on Mao's infallibility, which they invoked to justify their existence and ward off recriminations by local authorities. One result of the rebel assault on the party organization was the further concentration of power in Mao's hands. Ke Ming expressed this in a cogent metaphor: "All the small gods were overthrown—there was only one big god. Before, the party committee secretary had been a small god; not anymore." While the movement concentrated power at the top, it deconcentrated power at the bottom. Power passed from local party officials to mass organizations, which were competing for mass support, and students, workers, and peasants gained unprecedented—if temporary—power over officials who had previously dominated their lives.

Radicals Versus Moderates

In late December 1966, Kuai Dafu and Jinggangshan took charge of the Tsinghua University campus. By then, their Red Guard opponents had collapsed, and most student fighting groups had joined Jinggangshan. This unity, however, was short-lived. In February, Mao called on the mass organizations in schools, factories, and villages to form "revolutionary committees." A small team of military officers was dispatched to Tsinghua to persuade students to select leaders of their own fighting groups and a number of the old university cadres to serve on such a committee. Faced with the task of appointing a new university leadership team, students split into factions. Pleas by the military officers for conciliation were ignored, and the campus polarized into radical and moderate camps, with the radicals opposing the rehabilitation of all but a few university cadres, and the moderates demanding the rehabilitation of most cadres.[11]

The split was triggered by a March 1967 report published in *Red Flag*, the party's most authoritative journal, which criticized the work team sent to Tsinghua in the summer of 1966 for indiscriminately attacking university cadres.[12] The article reflected efforts by central party leaders to rein in the mass organizations and reestablish order, efforts that had—for the time being—gained Mao's acquiescence.[13] Leading members of Tsinghua's Jinggangshan organization responded with a big character poster that criticized the *Red Flag* article. The poster reaffirmed their position against rehabilitation, and claimed that the authors of the *Red Flag* article were trying to deceive Mao about the situation at Tsinghua. According to Ke Ming, a student party member who had supported Kuai's opposition to the work team, the debate about this big character poster was a turning point in the factional conflict at the university. "After that incident, people began to think about what the Cultural Revolution was really all about," he told me. "The central questions were: First, was Tsinghua under bourgeois or proletarian dictatorship? Second, were most cadres good or bad? Should they be overthrown or not?"[14] Ke Ming decided they should not. He joined other like-minded students in organizing a mass rally on April 14 to defend the "good cadres" at the university. The rally gave birth to the moderate faction. Because the new group emerged out of a split in Jinggangshan, it insisted on keeping the organization's name, but it was popularly known as "April 14."[15] After that, students, teachers, and workers at the university coalesced into two stable contending factions, one fighting for radical change and the other advocating moderation. Conflict between these two camps gripped Tsinghua for the next fifteen months.[16]

In the spring of 1967, schools, workplaces, and villages across China polarized into similar radical and moderate camps. Because of Tsinghua's stature and its proximity to the center of power, Jinggangshan became the most famous and influential radical organization in the country and the April 14 faction became a standard bearer of the moderate camp. Below I will compare the aims of the two organizations. In some ways, each represented its respective camp across the country, but there were also important differences between universities and secondary schools, factories, and villages, and the contending programs described below reflected the university context in which they were formulated.

UNIVERSITY RADICALS: ATTACKING POLITICAL AND CULTURAL CAPITAL

Kuai Dafu and those who remained in the radical camp enthusiastically took up Mao's entire Cultural Revolution agenda, attacking both the political and cultural status quo. The main tasks of the Cultural Revolution, Kuai wrote, were "to discredit and overthrow the authorities taking the capitalist road, to discredit and overthrow the bourgeois academic authorities, and to thoroughly reform the educational system and teaching methods."[17] As minister of higher education, Tsinghua's president, Jiang Nanxiang, had become a principal target of the Cultural Revolution, and Tsinghua, which had proudly called itself the "cradle of Red engineers," was now a prominent symbol of the dangers of the capitalist road. Both the academic and the political credentialing systems were now suspect, and the radicals' main slogan—"Thoroughly smash the old Tsinghua"—targeted both.

Fervent rhetoric against the political establishment was the radicals' hallmark, and their appeal was strongest among students who resented the party's tight political control, the power and privileges of Communist officials, and the system of career advancement based on political loyalty. The radicals fiercely criticized university officials and they enthusiastically attacked higher-level party leaders as well. "Our primary target was the capitalist roaders," Kuai told me. "Those who had already been overthrown—the so-called old Rightists, the old intellectuals, the old Nationalist Party—they were not the main problem. The danger of restoration came from within the Communist Party's own ranks, from some of its own leaders."[18] An article in Jinggangshan's newspaper declared, "Those taking the capitalist road have captured part of the state machinery in China (and it has become capitalist state machinery)." What was required, therefore, was "a great revolution in which one class overthrows another."[19]

The radicals directed their attacks not simply against individual leaders, but against fundamental features of the political system. They challenged the authority of the party's bureaucratic hierarchy, criticized its culture of political dependency, and denounced the system of career advancement based on political loyalty. Their goal, according to radical activists, was to do away with the existing "hierarchical system, cadre privileges, the slave mentality, the overlord style of work, and the bloated bureaucracy."[20] Their solution to all of these problems was organizing "mass supervision" over cadres, a task they took up with relish, dragging university officials up on stages to be criticized—and sometimes cruelly humiliated—by their subordinates. The greatest gain of the Cultural Revolution, Jinggangshan activists declared, was "destroying servile thinking."[21]

Radical efforts to condemn the party's culture of political dependency were given a boost by a campaign launched by the CCRSG in the spring of 1967 to criticize Liu Shaoqi's book, *How to Be a Good Communist*.[22] Liu, the party's organization chief, became the main target of the Cultural Revolution and he was subjected to hyperbolic criticism, including charges of treason that distorted party history. Nevertheless, as Lowell Dittmer made clear in his biography, Liu was a highly appropriate target for a campaign against bureaucratic authority.[23] He not only championed bureaucratic efficiency and order, but in his personal demeanor he modeled the discipline, deliberation, organizational loyalty, and domineering authority of a bureaucratic official. In Liu's book, which was required reading for those who aspired to join the party, he had stressed that Communists must submit to the will of the party organization. Mao initiated the campaign against the book, declaring, "Party members in the past were isolated from the masses because of the influence of *How to Be a Good Communist*, held no independent views, and served as subservient tools of the party organs. The masses in various areas will not welcome too quick a recovery of the structure of the party."[24]

Jinggangshan publicists used the campaign as an opening to attack the modus operandi of Tsinghua's party and Youth League organizations and, particularly, of their recruitment apparatus. They claimed that Party Secretary Jiang, like Liu, had encouraged careerism among party and league members, and had demanded subservience in exchange for promotion. They denounced Jiang's motto, "Be obedient and productive," and argued that he had cultivated a particularly servile group of cadres at Tsinghua. In a scathing essay published in the *Jinggangshan* newspaper, a middle-level university cadre wrote that Jiang's main criterion for selecting cadres was "obedience." The author, who described himself as a "pure Tsinghua-brand cadre," displayed a mastery of the criticism/self-criticism style required during the Cultural Revolution. "To be a good cadre, you had to obey 'Comrade Nanxiang' and the 'school party committee,'" he wrote. "As long as you were obedient, you could become an official, you were placed in an important position, and you were deeply grateful."[25] As a result of this kind of selection and of lengthy training at the university, the author continued, Tsinghua cadres had been particularly damaged by Liu's "self-cultivation" mentality. "They always stick to convention and have a slave mentality; in their work they are only responsible to those above them, and they care more about following the regulations than about right and wrong. While they are subservient yes-men towards those above them, they exercise a bourgeois dictatorship over those below them and suppress divergent opinions."[26]

While Jinggangshan activists scorned this "slave mentality," they took pride in their own "rebel spirit," by which they meant independent thinking and a willingness to challenge authority. "Those who thought creatively and had different opinions supported Jinggangshan," Liu Peizhi, a radical activist, told me. "I didn't care about the personal cost; if something was wrong—then challenge it."[27] Mao's axiom, "It's right to rebel," became the motto of the Cultural Revolution, and Jinggangshan activists celebrated a seditious and sometimes violent bravado that was more akin to the spirit of the Communist insurgency than to the orderly activism of the post-1949 party and Youth League organizations. Kuai Dafu, who would later spend seventeen years in confinement as a result of his prominent role in the Cultural Revolution, was fond of citing the traditional insurgent motto that was also a favorite of Mao's: "He who does not fear death by one thousand cuts dares to pull the emperor from his horse."

In opposing the political establishment, the stance of the radicals at Tsinghua University was in agreement with that of their younger allies at Tsinghua's attached middle school. Unlike their middle school counterparts, however, the university radicals had little sympathy for the existing education system and the old educated elite. The leadership and much of the membership of the radical faction at the university were from peasant and working-class families, and although the organization condemned the "theory of natural Redness" promoted by revolutionary cadres' children, it continued to stand by class line policies. While middle school radicals enthusiastically distributed Yu Luoke's article opposing the family origin system, few copies reached the university, and the university radicals condemned the article as a "big poisonous weed."[28] Liu Peizhi, a radical activist whose grandfather was a landlord, recalled that challenges to class line policies never gained much traction at the university. "People in [Jinggangshan] didn't have any sympathy for those opposing family origin. That kind of idea could not survive in China then—people still remembered the revolution against the landlords, so it just couldn't get much support."[29]

Jinggangshan condemned Tsinghua's "revisionist education line" for reproducing class differences. The national entrance examinations and university admissions policies, the radicals argued, favored students from old elite families and from large urban areas. Even in big cities, workers' children were at a disadvantage. Among students recruited from Shanghai, *Jinggangshan* commentators noted, the number of students of bourgeois origin was twice the number of working-class origin.[30] As a remedy, a Jinggangshan committee established to study the recruitment problem suggested sweeping changes in the admissions process. Sixty percent of students, they wrote, should be selected through a process of recommendation by the masses, and entrance exams should be thoroughly reformed and only reintroduced in a supplemental role. Children of worker and peasant origin should make up 65 percent of all students admitted, and no more than 5 percent should come from the former exploiting classes.[31]

Tsinghua, Jinggangshan leaders claimed, had been turned into a "breeding ground for capitalist successors." By recruiting students from privileged families, encouraging the isolation of students from the masses of workers and peasants, denigrating productive labor, and neglecting political education, they charged, the university was cultivating an "intellectual aristocracy." If elitist education practices were not changed, they would inevitably generate a class hierarchy. "The 'superior' would get more 'superior' and the 'inferior' would get more 'inferior,'" wrote the radicals, leading to a situation in which the "superior" would become an exploiting class "standing on the heads of the 'inferior' working people."[32]

Students not only had to destroy the old Tsinghua, radicals argued, but they also had to transform themselves. "The students were tied intimately to the old Tsinghua. . . . We had been trained as this kind of intellectual aristocrats," Kuai Dafu told me. Mao wanted students to "completely negate (*fouding*) themselves," he explained. "He wanted us students to be common people (*pingmin*); he wanted us to unite with the workers and peasants. . . . On this point we really believed Mao Zedong. We saw we had become isolated from the workers and peasants, from the people, and that we should go back to the workers and peasants and unite with them."[33]

UNIVERSITY MODERATES: DEFENDING POLITICAL AND CULTURAL CAPITAL

Those who rallied behind the April 14 banner opposed Jinggangshan's wholesale condemnation of the leadership and policies of the old Tsinghua. During the seventeen years before the Cultural Revolution, they argued, mistakes may have been made, but the dominant line was always correct.[34] Moderate leaders urged students to "courageously protect, enthusiastically help, and boldly employ" cadres who had made mistakes. "They have rich experience in revolutionary struggle, they have the ability to hold power for a proletarian country and understand how to manage and run academic activities," the moderates declared in a front-page editorial in one of the first issues of their newspaper. "We must have the proletarian revolutionary guts to defend them." Former upper-level cadres, they argued, should be allowed to join the Cultural Revolution and, in fact, should become the backbone of the movement. With individual exceptions, even top leaders deemed to be capitalist roaders should be rehabilitated. "We should drag them out from under their beds, give them work, and let them make up for their mistakes."[35] The former leaders, they contended, were more capable of running the university than radical students who only knew how to shout about tearing things down.

April 14 leaders opposed Jinggangshan's contention that a new privileged stratum was emerging in China, arguing instead that "class relations had remained stable" since 1949. The capitalist roaders in the party were a small group, moderate leaders argued, and they were dangerous not because they represented a new privileged stratum, but rather because they represented the old exploiting classes. Jinggangshan's "ultra-Leftist" talk about a new privileged stratum and about one class overthrowing another would only open the back door for the old exploiting classes to come back to power.[36] "Our aim is to strengthen the leadership of the party, not weaken it," moderate students argued in a big character poster. "The theories about 'thoroughly reforming the dictatorship of the proletariat,' 'thoroughly opposing everything from the past,' 'thoroughly overthrowing everything from the past,' and 'rebuilding the party from scratch after the Cultural Revolution is over,' etc., are totally wrong."[37] Although the radical climate compelled moderates to criticize a few top party leaders, they tried to narrow the targets. "Then you couldn't be too moderate—if you were, people would call you conservative," said Zhu Yongde, explaining why he and his comrades in the April 14 faction criticized Jiang Nanxiang, even though they had positive feelings for him. "But we supported the rest of the cadres—the great majority of the cadres. We had our views, but we were limited in what we could say openly, so Shen Ruhuai [the top April 14 leader] promoted the idea of liberating the cadres, starting with the lower levels."[38]

The moderates' newspaper condemned the "revisionist education line" of the previous seventeen years, but its defense of the Tsinghua establishment clearly relayed another underling message. In contrast to Jinggangshan's wholesale condemnation of the old Tsinghua, the April 14 faction insisted that although the university's education policies were wrong, it was necessary to recognize that "one divides into two," that is, that not everything was bad. "We didn't dare say the old education system was good, but we thought part was good and you should reform the other part," said Ma Yaozu, a moderate leader. "You shouldn't just overthrow everything. The old intellectuals and cadres have a role to play. . . . So we protected the cadres, the teachers, and the intellectuals."[39]

CLASH OVER THE "NEW BOURGEOIS INTELLECTUALS"

From the radicals' perspective, the group that represented the greatest danger was the "new bourgeois intellectuals," that is, those who had received university training since 1949. They were more dangerous than the "old bourgeois intellectuals"—those who had been trained before 1949—because the latter had little political power. According to the radicals, the new generation of bourgeois intellectuals were still mainly from nonlaboring families and as a result of Tsinghua's elitist education policies they also maintained a bourgeois worldview; moreover, many of them also enjoyed the legitimacy and power associated with the party. "Most of them have the outside appearance of being 'Red and expert,' some are also party members and have cadre titles, so they are adept at misusing the party's name to promote revisionist garbage," an article in the radical newspaper argued. "They have political capital and prestige and most things have to go through them to get done."[40] Thus, this new generation of intellectuals was of particular concern to the radicals because they had both cultural and political capital. They were, Jinggangshan leaders charged, the main social base of the Jiang administration, and they had become part of a "privileged stratum" (*tequan jieceng*) at the university, which also included top university officials and sections of the old bourgeois intellectuals. The university party committee, they wrote, "was certainly not the vanguard of the proletariat, but rather the agent of the new bourgeois intellectuals."[41]

When it came to rehabilitating university cadres, Jinggangshan leaders would only agree to bring back those who were of worker-peasant origin and did not have university degrees. They were particularly opposed to cadres who had graduated from Tsinghua—the so-called Tsinghua-brand cadres. Li Guangyou, a Jinggangshan activist, said his faction distrusted these cadres because they were intellectuals. "Tsinghua University graduates—those people were no good. . . . They implemented Liu Shaoqi's capitalist line," Li told me. "Intellectuals are a minority—a very small minority. Our faction believed that even if they are in the Communist Party, even if they are party cadres and ought to represent the workers, there was no way they could represent the workers and peasants. Only worker-peasant origin cadres could represent them."[42]

The April 14 organization fiercely disagreed with Jinggangshan's "new bourgeois intellectual theory."[43] Those who had been educated after 1949 could not be considered bourgeois intellectuals, they argued, because they had been trained under the Communist Party. The moderates particularly objected to attaching the adjective "bourgeois" to university graduates who had joined the party. Ninety percent of university cadres were this type of Tsinghua-brand cadre, the moderates argued, and the great majority were good and should be rehabilitated.[44]

The debate over the cadre problem came to focus on one highly successful young university leader, Lu Yingzhong. Born into a well-to-do family, Lu was among the Tsinghua students who embraced the Communist cause on the eve of the 1949 Revolution. He studied under Qian Weichang and after graduating with a degree in mechanical engineering in 1950 he was hired by the university and gained renown as one of the most technically proficient of the young Communist cadres. By 1958, Jiang Nanxiang was presenting Lu as a model of the kind of Red engineer that Tsinghua aimed to train, and he called on students to follow the "Lu Yingzhong road." In 1960, Lu was asked to lead a team of one hundred students in designing and building a nuclear research reactor, which was completed in 1964. The debate in 1967 was not so much about the man as about the road named in his honor. For Jinggangshan, this was the capitalist road, which was producing new bourgeois intellectuals who were becoming a privileged stratum. For the April 14 organization, Lu represented the best of the postrevolutionary generation of Communist leaders, a stalwart of the new socialist order. In his memoir, Shen Ruhuai, the top moderate leader, explained why he and his organization decided it was critically important to stand behind Lu. "Lu Yingzhong was a member of the standing committee of the Tsinghua University party committee, he was the head of the nuclear research reactor and the secretary of its party branch, and he was a cadre who had been trained by Tsinghua University itself after liberation," Shen wrote. "He was not only a leadership cadre, but he was also a scientist. We were justly proud of his achievements. . . . The Lu Yingzhong road was the 'Red engineer' road, it was the 'Red and expert' road. Up through the Cultural Revolution, I always considered Lu Yingzhong a model that we should emulate."[45]

Leadership and Membership of the Radical and Moderate Factions

In contrast to the contending factions at Tsinghua's attached middle school, which were led by children of revolutionary cadres and intellectuals, at Tsinghua University the principal leaders of both factions were students from peasant and worker families. The backgrounds of the two leading adversaries, Kuai Dafu of the radicals and Shen Ruhuai of the moderates, were very similar. Like Kuai, Shen came from a peasant family and he had also been an accomplished activist in the Tsinghua party and league organizations before the

Cultural Revolution. Shen's parents were not party members (as Kuai's were), but they both had been activists in the mutual help movement that laid the groundwork for collectivization. After a school was opened in his village in 1952, Shen became a diligent and capable student and he tested into Tsinghua in 1965. He joined the party soon afterward and was selected as leader of the party group and secretary of the league branch in his class.[46] Although Shen and Kuai shared similar social origins, this did not translate into a common political orientation during the Cultural Revolution. "My viewpoint then was very clear," Shen wrote in his memoir. "The school party committee was the representative of the party at the school; you could only defend it, you could not oppose it, because it was the liberator of the Chinese people." The party had rescued his family from destitution and it had given him—a village youth—the opportunity to study at the university, so he was deeply grateful. "My entire life would not have been possible," he explained, "if it were not for the party."[47] Kuai also professed deep feelings for the Communist Party (which he, too, attributed to his rural origin), and, like Shen, he was acutely aware of the social distance that separated the lofty world at Tsinghua from the village in which he had grown up. "I came from a very poor village," he told me. "Tsinghua seemed like paradise." Kuai, however, was not satisfied with the existing order. "Then everyone said they supported socialism, but we were the ones who really supported socialism, . . . the truest, most thoroughly revolutionary socialism. . . . We supported the socialism that Mao Zedong demanded. . . . No exploitation, no inequality, the workers and peasants should truly become masters of society."[48]

Of the nine most important leaders of Jinggangshan, eight were of worker or peasant origin; on the April 14 side, six of the nine most important leaders were of worker or peasant origin. Cultural Revolution norms dictated that students of "good" class origin take the lead, and in the spring of 1967, students of worker or peasant origin were in a better political position to claim leadership positions than students from revolutionary cadre families. By then, Liu Tao, He Pengfei, and other children of high-level cadres had dropped out of political activity and their Red Guard organization had disbanded. Their parents were now under attack and their pedigrees had become an acute liability. Many students whose parents were lower-level revolutionary cadres continued to participate, and several became factional leaders, but they were also vulnerable. Even more vulnerable were students from old elite families. As a result, the most prominent leaders of both factions were children of workers and peasants.[49]

Unlike at Tsinghua's attached middle school, where children of the political elite lined up on one side and children of the educated elite on the other, at Tsinghua University students never divided along family origin lines. Even in the summer of 1966, when bloodline theory was ascendant, the social composition of the two factions at the university—one supporting and the other opposing the work team—was mixed. Although the pro–work team faction loudly advocated bloodline theory and proclaimed itself to be the champion of the "five Red classes" (revolutionary cadres, revolutionary soldiers, revolutionary martyrs, poor and lower-middle peasants, and workers), many students of peasant and worker origin were not impressed and instead joined the anti–work team camp. The subsequent triumph of Jinggangshan and then its split into radical and moderate camps completely jumbled the original factional affiliations, and the split produced organizations that were even more mixed in terms of family origin.

The extent of the factional realignment in the spring of 1967 can be seen in Table 5.1, which reproduces the results of a campus survey conducted shortly after the split. While the leaders and many of the activists in the new moderate organization had been part of the anti–work team camp, they were now joined by the majority of those who had supported the work team, and while most of those in the radical camp had been anti–work team rebels, they now also counted among their ranks a substantial number of former supporters of the work team. Moreover, individuals who had not participated in either faction during the early months of the Cultural Revolution (many because of problematic family origins) also ended up on both sides.

TABLE 5.1
Spring 1967 factional realignment at Tsinghua University

	Radicals				Moderates			
	Anti–work team	Pro–work team	Neither	Total	Anti–work team	Pro–work team	Neither	Total
Totals	1,011	534	588	2,133	1,486	1,815	320	3,621
Percent	47	25	28	100	41	50	9	100

SOURCE: Figures are derived from data in a table published in Shen (2004, 116). This data was originally from an internal report produced by the Jinggangshan Regiment on April 26, 1967. I revised one figure that seemed to be the result of faulty addition.

Everyone I interviewed who was at Tsinghua University during the Cultural Revolution insisted that university students did not split along "good" and "bad" family origin lines. Both the radical and moderate factions, they reported, had many members of peasant and worker origin, and they also had many members from urban educated families. Interviewees generally agreed that students from revolutionary cadre families were more likely to support the moderates, and students of "bad" family origin were more likely to support the radicals (if they participated at all), but they also reported that students from both groups could be found in both factions. When I asked interviewees what distinguished students who supported the moderate camp from those who supported the radical camp, the most common answer was that those with stronger connections to the party organization tended to support the moderates. This understanding was confirmed by data gathered in the spring 1967 survey: out of 1,631 student cadres and party members, 1,029 (63 percent) joined the moderate faction.[50] Nevertheless, the radical camp also included many students, like Kuai Dafu, who had been very active in the party and the Youth League.

When I asked Kuai whether the family origins of students in the two factions were different, he replied with a simple class analysis. "More of those from the lower levels—children of common people—supported us," he told me. "Those from higher levels—middle-level and above cadres, high-level white-collar employees, including those from Tsinghua—their children supported them. Generally speaking, those from families who had a little more education, who had higher education, they tended to support them."[51] He admitted, however, that many students of worker and peasant origin fought on the moderate side, and that his own faction included students from highly educated and politically connected families. Shen Ruhuai, the leader of the moderate faction, claimed in his memoir that the majority of those who were of "good" family origin and had been politically advanced before the Cultural Revolution were on the moderate side, although he added that neither family origin nor political standing completely determined factional affiliation.[52] Zhou Quanying, the most famous polemicist in the April 14 camp, stressed political affiliation rather than family origin in explaining the differences between the two factions. In his memoir, Zhou suggested that his faction represented the interests and political inclinations of the "advanced people," including political counselors, student cadres, and party members, while the radical faction represented the "common people," that is, students, teachers, and employees who had not joined the party or been active in its political activities.[53] Another moderate leader linked the "extreme" stands taken by the radicals to the social position of its membership, explaining that many radical activists came from "bad" families or from very poor peasant households. The moderate stand of his own faction, he added, was linked to the "middle class" position of the bulk of its membership. Although these four descriptions are somewhat contradictory, they each represent a vague consensus understanding: the moderates represented those who were more firmly connected to the established order, while the radicals represented those whose ties were more tenuous.

Although top university officials could not participate in the movement, many other university employees—including middle- and lower-level cadres, teachers, staff, and workers—supported one faction or the other. Table 5.2 presents data from a survey conducted in May 1967 about factional alignment among Tsinghua University employees. As was the case with students, involvement with the university party organization was an important factor;

TABLE 5.2
Factional alignment among Tsinghua University employees

		Radicals		Moderates		Neither	
	Total	Number	%	Number	%	Number	%
All university employees	3,267	1,324	41	999	31	944	29
Professors (senior faculty)	151	63	42	22	15	66	44
Workers	994	537	54	205	21	252	25
All other university employees (principally junior faculty)	2,122	727	34	772	36	626	30
Party members (among all university employees)	1,113	336	30	528	47	249	22
Nonparty members (among all university employees)	2,154	988	46	471	22	695	32

SOURCE: Shen (2004, 117–18). The survey did not include all university employees. In 1964 there were 5,566 university employees, of whom 1,718 were workers and 2,226 were teachers, including 191 professors (Fang and Zhang 2001, vol. 1, 482, 490).

the moderates had more support among employees who were party members, and the radicals had more support among those who were not. Although both factions had support among all sectors of university employees, a pattern of factional alignment emerges that is similar to that among the students. The moderates were stronger at the center of the university establishment, a space defined by both Redness and expertise, and occupied first and foremost by young teachers. The radicals, in contrast, counted more on people at the margins, among those who were expert but not Red (older faculty), Red but not expert (worker-peasant cadres), or neither Red nor expert (many of the workers).

Middle-level university officials played a prominent role in the formation of the moderate faction, and soon after the April 14 rally 150 middle-level cadres signed an open letter supporting the new organization. Moderate student activists described university cadres as "backbone elements" in their organization, and Qiu Maosheng, a moderate leader, proudly told me, "Most lower and middle-level cadres either participated in April 14 or sympathized with us."[54] Luo Zhengqi, deputy head of propaganda for the Tsinghua party committee before the Cultural Revolution, and Tan Haoqiang, deputy head of the school committee of the Youth League, helped write several of the moderates' key political statements. Tan was invited to join the organization's top leadership committee, as were two teachers who were leaders of faculty fighting groups.[55]

Because Jinggangshan staked out a position hostile toward Tsinghua's political establishment, it is not surprising that it had less support among university cadres. Several top university officials formally endorsed the organization, but radical and moderate activists alike considered these endorsements no more than calculated moves to mitigate criticism as "capitalist roaders." There were a handful of important university- and department-level cadres whom Jinggangshan activists considered to have genuinely repudiated the "revisionist education line" and accepted the ideas of the Cultural Revolution, but even these cadres were kept at a distance. The only cadre invited to join Jinggangshan's leadership committee was Han Yinshan, a veteran revolutionary of peasant origin, who had lost his position on the school party committee in the 1950s. Jinggangshan also enjoyed the support of a particularly radical group of faculty members, the Red Teachers Union, some of whom reportedly had problematic family histories and all of whom were outspoken in their opposition to the political system (some called the university party organization "a black fascist party").[56]

Uniting to Defend the Status Quo

The moderate camp that emerged in the spring of 1967 is often described as a continuation of a conservative tendency first represented by the pro–work team Red Guards in the summer of 1966.[57] At Tsinghua, however, the political orientations of the two organizations were very different. The original Red Guards defended political capital, but they attacked cultural capital. They defended the work team, which represented higher party authorities, but they violently attacked the old Tsinghua including the entire university party organization, which they associated with the "revisionist education line." The April 14 coalition, in contrast, became the de facto champion of the old Tsinghua and the university party organization. The original Red Guards were known as the organization of the "high cadres' kids" (*gaogan zidi*) and their extreme class line rhetoric alienated the majority of Tsinghua students, who were from the old educated elite. In contrast, the April 14 coalition was a conservative political alliance that included children of both revolutionary cadres and intellectuals. Students who had been on opposite sides of the barricades in the summer of 1966 were now comrades in arms in the April 14 organization: according to the survey results presented in Table 5.1, 50 percent of the members of the April 14 faction had supported the work team and 41 percent had opposed it. They continued to have different perspectives, but they came together to defend the status quo against radical attacks. Both the differences among moderates and the common ground that brought them together can be seen in the stories told by the following three individuals.

The first was an anonymous former member of the original Red Guards who submitted a personal statement to the moderate newspaper in July 1967. Because he came from a revolutionary soldier family, he wrote, he loved Chairman Mao and the Communist Party from the depths of his heart. After he arrived at Tsinghua, however, he was disturbed to find an atmosphere that emphasized performance instead of class background and he thought many of the university's practices—in admissions, party and league recruitment, and selection for the "teach according to ability" program—were contrary to the party's class line. As a leader of his Youth League branch, he raised these issues with university officials, but they ignored his concerns. In June 1966, when he heard that Jiang Nanxiang was being criticized, he was very excited and he cried with joy when the work team arrived. As a member of the Cultural Revolution preparatory committee appointed by the work team, he supported the "bourgeois reactionary line," but he later renounced it. Now a year later, he wrote, he was very happy to be able to join the April 14 faction. He felt it was unfair, however, that some members of the organization treated him with suspicion; he argued that former Red Guard activists like him should be treated as "backbone" members of the organization.[58]

The second, a student named Li Weizhang whose father was a teacher in Wuhan, had opposed the work team in the summer of 1966. Before the Cultural Revolution he had won a leadership position in the Youth League in part because of his academic abilities. He greatly admired his professors and President Jiang, and he was shocked when Jiang was removed by the work team. He was further alienated from the work team—and its Red Guard supporters—because of their antidemocratic methods and their stress on class background. In the summer and fall of 1966, Li sympathized with Kuai Dafu because he championed "big democracy" (*da minzhu*). Kuai's "extreme class line," however, increasingly alienated Li, and he gravitated to the April 14 faction because it was "not too extreme against the school officials." The moderate faction, he told me, was the natural home for students like him, who did well academically, had been Youth League leaders, and had good relationships with the department leadership. "I think normally that people that used to feel comfortable [with Jiang] before the Cultural Revolution naturally tended to the April 14 faction. He said, 'Everyone is equal under grades,' so that's why people with my background felt comfortable with that emotionally."[59] In Li's understanding, possession of political capital (leadership in the Youth League and a good relationship with the department leadership) led students to support the April 14 faction, but possession of cultural capital also motivated support for the moderate camp, which was seen as more amenable to preserving the old Tsinghua's meritocratic policies.

Wang Jiahong, another student of intellectual origin whose political credentials were much weaker than Li's, confirmed this. His parents were both professors at Peking University and both endured harrowing criticism during the Cultural Revolution. His father's Harvard education cast a shadow over Wang's own future, and he was the last student in his class to be admitted into the Youth League. In 1966, he was an even more enthusiastic rebel than Li, relishing the chance to challenge the political system. Nevertheless, Wang also joined the moderate faction in 1967. The main difference between the two factions, he figured, was that the moderates would preserve more of Jiang Nanxiang's education policies. "[Jiang's] class line was more moderate than the other because he had been dealing with educated people—intellectuals—and he knew in order to achieve something academically or in the economy or construction you need knowledge and a lot of the people at that time who possessed the knowledge . . . were not from the correct background," he told me. "In his policy, if you are academically outstanding and your family was not bad enough not to dare to admit you, then when he took you in he would give you the proper conditions for you to academically achieve something."[60] Jinggangshan, Wang recalled, "took a much more radical attitude toward everything Jiang Nanxiang had said prior to the Cultural Revolution. Their slogan was 'Smash to the ground everything of the old Tsinghua.'" Wang, therefore, joined the moderates to defend the old Tsinghua. Although he resented Jiang's rigid political control and the class line discrimination he had encountered, he found Jiang's meritocratic policies more palatable than Jinggangshan's wholesale attacks on Tsinghua's education policies.

From the testimony of these three moderate activists, it is clear that there were still sharp points of conflict between children of the new and old elites. The son of the revolutionary military officer was keener to defend political capital, and much less inclined to defend cultural capital, while the opposite was true of the teachers' sons. Nevertheless, they all came from relatively comfortable homes, they all did well enough academically to test into Tsinghua, and they were all members of the Youth League, all of which helped dispose them to defend the status quo. In the factional alignment that developed at Tsinghua University in the spring of 1967, to defend cultural capital was to defend political capital, and vice versa. Both had been pillars of the old Tsinghua that the moderate faction was committed to defending.

The Cultural Revolution and the Coalescence of Political and Cultural Elites

The prevailing explanation of student factional conflict during the Cultural Revolution highlights contention between children of party officials and children of intellectuals. This explanation, which might be called the competing elites model, was most cogently presented by Anita Chan, Stanley Rosen, and Jonathan Unger in their investigations of student factions at elite Guangzhou middle schools.[61] The great majority of the students in these schools came from revolutionary cadre or intellectual families, and the central rift between them was produced by class line policies. "Conservative" factions, led by children of revolutionary cadres, promoted class line policies, focused their attacks on teachers, education officials, and members of the old educated elite, and defended the authority of the party, while "rebel" factions, led by children of intellectuals, denounced class line policies and focused their attacks on political authorities. The competing elites model describes well the factional alignment at Tsinghua's attached middle school, but it does not capture the nature of the conflict at Tsinghua University. Table 5.3 presents a simple diagram of the orientations of the principal factions at Tsinghua University and its attached middle school in relation to political and cultural capital. At the middle school, the original Red Guards attacked cultural capital and defended political capital, while the rebels attacked political capital and defended cultural capital. At the university, in contrast, the radical faction attacked both political and cultural capital, while the moderate faction defended both.

An important cause of the different factional alignments at the two schools was demographic. Tsinghua's attached middle school was an elite urban middle school in which the great majority of the students were from intellectual or revolutionary cadre families. When Mao encouraged attacks on the political and educated elites in 1966, existing fractures between students at the middle school were split wide open. Each of the two factions ardently supported one aspect of Mao's agenda, with revolutionary cadres' children spearheading the attack on cultural capital, and intellectuals' children leading the assault on political capital. By the same token, neither of the two factions could embrace Mao's entire program. An organization made up almost entirely of cadres' children could not join the assault on political capital, while an organization composed almost entirely of intellectuals' children could not attack cultural capital. Bloodline theory produced factions so strongly defined by elite family

TABLE 5.3
Cultural Revolution factions at Tsinghua University and attached middle school

Axis of contention		Political capital	
		Attack	Defend
Cultural capital	Attack	University radicals	Middle school Red Guards
	Defend	Middle school rebels	University moderates

origin that neither could violate principles so dear to their collective interests and identities. Moreover, because the school was so firmly split along family origin lines, there was also little room for a group that promoted conciliation between the revolutionary cadres' children and intellectuals' children to defend the status quo. Although a moderate faction that included both children of revolutionary cadres and of intellectuals did emerge, it was always feeble.

At Tsinghua University, in contrast, nearly 40 percent of the students came from peasant or worker families. This prevented the polarization of the campus into two camps based on elite identities. Although, at the outset of the upheaval, revolutionary cadres' children also promoted bloodline politics, the theory was less potent at the university because students from worker and peasant families were neither completely accepted nor completely excluded by its propositions. The two factions that ultimately emerged could not be easily distinguished in terms of family background and they were both led by students of worker or peasant origin. This meant that neither faction was bound to the collective interests and identities of either the political or the intellectual elite. The radical faction, led by children of workers or peasants and supported by students of diverse social origins, fully embraced Mao's call to attack both political and cultural capital. Their attacks called into being an adversary that defended both, creating an alliance between political and cultural capital. Facing the radical challenge, many students—including children of both the political and the intellectual elites—found grounds to unite in defense of the status quo.

The competing elites model has been widely interpreted by academics in the West as a general model for student factional conflict during the Cultural Revolution. It is often overlooked that Chan, Rosen, and Unger derived their explanation from studying one type of school—elite urban middle schools—and that they warned against unqualified generalizations. Although they chose to focus on the highly polarized pattern at elite middle schools, they noted that they found divergent patterns at other schools. At ordinary and vocational middle schools, they found the most salient division was between those who had been active in the Youth League and those who had not, and Rosen noted that factional alignment at Guangzhou universities also seemed to follow this pattern.[62] It seems reasonable to suggest that the competing elites model works well in elite urban middle schools, where the great majority of students were from either intellectual or revolutionary cadre families, but not at universities, colleges, and ordinary or technical middle schools, where there were substantial numbers of students of nonelite origin. It is likely that the factions in these schools looked more like the factions at Tsinghua University than those at Tsinghua's attached middle school. Indeed, Tang Shaojie, and Song Yongyi and Sun Dajin have identified the radical and moderate factions at Tsinghua University as representative of similar factions in other universities, and at the time the two were considered standard bearers of opposing camps nationwide.[63]

The battle lines at Tsinghua University reflected a new political reality produced by the Cultural Revolution. While the factional conflict at Tsinghua's attached middle school reproduced the battle lines of 1957, when intellectuals contended for power with party officials, the conflict at Tsinghua University transcended these old inter-elite conflicts and reflected the extent to which political and cultural capital had coalesced. In launching the Cultural Revolution, Mao's aim was to mobilize a popular movement against both the political and cultural elites. Although the initial result was an eruption of inter-elite conflict, the movement ended up forging unity between the two groups. In the factional conflicts at Tsinghua University and its attached middle school it is possible to see both phenomena. On the one hand, the rival camps created by bloodline politics at Tsinghua's attached middle school were a dramatic manifestation of inter-elite animosity; on the other hand, the moderate coalition that emerged at Tsinghua University represented a tenuous coalition of political and cultural capital.

The Cultural Revolution can be seen as a decisive moment in this process of inter-elite conciliation. The moderate coalition at Tsinghua University presaged the political unity between China's political and cultural elites, which would be consummated a decade later after Mao died and the Cultural Revolution was repudiated. In the meantime, the freewheeling factional conflict of the first years of the Cultural Revolution gave way to harsh institutional measures designed to undermine the power of the political and educated elites and to inhibit their convergence. These measures, as they were implemented at Tsinghua, will be treated in the next three chapters.

PART THREE

Institutionalizing the Cultural Revolution (1968–1976)

Chapter Six

Supervising the Red Engineers

On the morning of July 27, 1968, more than thirty thousand workers from factories around Beijing arrived at the gates of Tsinghua University. By that time, the radical and moderate student factions had been engaged in a violent contest for control of the school for over three months, and similar battles were taking place across China. Violence was escalating despite Central Committee directives calling on mass organizations to cease fighting. The workers were mobilized to descend on Tsinghua to enforce these directives. The moderate faction at the university, which favored a return to normalcy, welcomed the workers, but the radicals were determined to resist what they saw as an attempt to throttle the Cultural Revolution. They tried to hold the buildings they occupied, using spears and rifles to battle the legions of unarmed workers. By the next morning, the workers had won control of the campus, but five of their comrades lay dead and hundreds had been wounded.[1] In the early hours of that morning, Mao Zedong summoned Kuai Dafu and four other prominent leaders of university factions in Beijing to a meeting, which was also attended by Zhou Enlai, Lin Biao, and members of the CCRSG. At the meeting, Kuai reported that workers "manipulated by a black hand" had entered Tsinghua to suppress the students. Mao replied, "The black hand is me."[2]

A few days later, a Workers Mao Zedong Thought Propaganda Team composed of several thousand workers and soldiers was dispatched to take control of the university. The arrival of the workers' propaganda team, as it was called in shorthand form, was heralded in the national press as the first step in establishing a new era in which the working class would take charge of education, and the moment was consecrated with a basket of mangoes, Mao's personal gift to the team. The workers' propaganda team was instructed to supervise Tsinghua University on a permanent basis, and it became a model for similar teams sent to schools around the country.

The dramatic end of the "hundred day war" at Tsinghua was a watershed event that ushered in the systematic suppression of factional fighting throughout China and marked the end of the first period of the Cultural Revolution. Mao abandoned the rebel organizations he had called into being, but continued to pursue the goal that had inspired him to launch the Cultural Revolution—preventing the consolidation of a new privileged class. He was convinced that both the party organization and the school system were cultivating elite groups that set themselves apart from the masses, and he continued to be preoccupied with these concerns until his final days. After party members became high officials, Mao complained in a statement dictated shortly before his death, "they want to protect high officials' interests; they have a fine house, a car, high salaries, and even servants, they're even worse than the capitalists." After students graduate from the university, he added, "they think they're better than workers and peasants; they want to be a labor aristocracy."[3]

The party organization and the education system, which had been the main institutional targets of the Cultural Revolution and had largely ceased to function during the previous two years, were now rebuilt. In fact, they both grew substantially between 1968 and Mao's death in 1976, but reconstruction was carried out within the constraints of Mao's radical agenda. These years, often called the late Cultural Revolution, were a period of harsh and disruptive institutional experiments designed to check the capacities of the education system and the party organization to generate class differences. To press his agenda, Mao cultivated a coterie of radicals led by Jiang Qing, Zhang Chunqiao, Yao Wenyuan and Wang Hongwen, who after their downfall in 1976 would be known derisively as the Gang of Four. Jiang, Zhang, and Yao had been members of the CCRSG (which was disbanded in 1969), and Wang was the leader of the main rebel workers' organization in Shanghai. Mao's radical lieutenants maintained a factional network within the party composed largely of leaders of disbanded radical mass organizations who had been brought into positions of power. Although divisions and alliances at every level were complicated, Chinese politics during this period was largely animated by the polarization of the party into radical and conservative camps.[4] Deng Xiaoping became the standard bearer of the conservative camp, whose main base of support was made up of veteran cadres. While the radicals championed policies associated with the Cultural Revolution, the conservatives sought to constrain or reverse these policies.[5]

The main battle call of the radical faction was to "prevent capitalist restoration." Among the most elaborate treatises on this subject were those produced at Tsinghua University and Peking University, both of which were placed under radical control. Using Mao's language, the radicals named their enemy a "bureaucratic class," a term that was appropriate because they were targeting party and state officials and they were concerned, first and foremost, about the political institutions from which these officials drew their power. The term, however, fails to adequately convey their simultaneous concern about the increasingly technocratic character of these officials. Taking note of the Soviet experience, radical theorists specifically targeted Communist officials who had advanced technical credentials, the Red experts, as the vanguard of the peaceful counterrevolution, and they saw the education system as a key culprit. The Soviet education system, according to a widely publicized article penned by a writing group at Tsinghua and Peking universities, had trained the privileged stratum that was responsible for restoring capitalism. "We know that most of the heads of the Soviet revisionist renegade clique are so-called Red experts who received university training; they were precisely the ones who transformed the first socialist state into a social-imperialist state."[6] The radical policies of the late years of the Cultural Revolution were designed to undermine the institutional foundations of the emerging stratum of Red experts in China. Communist leaders had always had an ambivalent attitude toward the education system, which they saw as the domain of the educated classes, and efforts during the late years of the Cultural Revolution to overhaul this system were the continuation of a long-standing Communist agenda, although more drastic means were now employed.

The radical agenda in the political field was also not completely unfamiliar, as Mao had previously initiated numerous campaigns against bureaucracy, but the radicals' crusade against the bureaucratic class during the late years of the Cultural Revolution was in many ways unprecedented. While Communist theorists had previously identified the threat of "capitalist restoration" as coming primarily from outside the country and from the old propertied classes, radical writers now located the main danger as arising from party officials and structural defects in the socialist system. "Capitalist roaders are a product of the historical period of socialism," declared the author of an article in the official Tsinghua University newspaper. "It is inevitable that the conditions that give rise to them will be here for a long time. After the proletariat takes power, the result is state capitalism without the bourgeoisie. . . . [The capitalist roaders] want to expand . . . the power the people have given them, and turn it into the privileged right (*tequan*) to rule over and oppress the people."[7] An article in the country's leading newspaper, *People's Daily*, used even more inflammatory language to drive home the same message. "These people revel in high positions, they sit on the heads of the masses of workers and peasants, indulge in comfort, and abuse their power," the author wrote. "As they see it, . . . it is completely natural for officials to have the final say in everything. As a matter of fact, they place themselves in the same position as the big bosses of the old society and treat the masses of workers and peasants as employees or slaves."[8]

The essential problem, in the radicals' view, was the concentration of power in the hands of Communist cadres and the exclusion of the masses of workers and peasants from decision making. Members of the bureaucratic class, a radical theorist wrote in an essay published in a Peking University journal, "transform their authority into capital. . . . Once they have power, they always try to turn their professional authority into privilege, and turn the management and service authority the people gave them into power to repress the people and deprive them of power. . . . They do everything they can to exclude the laboring people from management and planning work, and to take into their own hands the right to manage and distribute the means of production and the wealth produced by labor, thus eliminating in all but name the socialist public ownership system." During the socialist era, the author continued, "there is still a very fixed division of labor between leaders and the led, and between managerial and technical employees and the direct producers," and in work units practicing a "revisionist managerial line" these divisions were being reinforced and even expanded. According to radical doctrine, the consolidation of a new exploiting class could only be avoided by eliminating the division between mental and manual labor, promoting the participation of the masses in management, preventing cadres from becoming divorced from labor and from the workers and peasants, and inhibiting them from seeking privileges.[9] Their goal was to disperse possession of decision-making authority and knowledge, both of which were concentrated in the hands of managerial and technical cadres, and enhance the power of the masses.

The radicals, whose outlook was shaped by the battles of the early years of the Cultural Revolution, fundamentally distrusted the party's bureaucratic apparatus, which in their view was becoming an entrenched mechanism of class power and privilege. They were determined to create institutional means to check the power of party and state offices, to enable supervision of Communist cadres from below, and to weaken patterns of political dependence based on the party organization. Tsinghua University became a radical bastion, and the leaders of the workers' propaganda team at the university were closely associated with the radical camp. The team's principal leaders were two young military officers, Chi Qun and Xie Jingyi, both of whom had close ties to Mao Zedong and his radical followers. Chi had been deputy leader of the propaganda department of Division 8431, the unit that protected Zhongnanhai, the central party and state headquarters.[10] Xie had served as one of Mao's personal secretaries for a decade and she had close ties with his wife, Jiang Qing. Chi and Xie both played key roles in national politics. In 1970, Chi was appointed deputy leader of the State Council's Science and Education Group, and Xie became a deputy secretary of the Beijing Municipal Party Committee and was elected to the party Central Committee in 1973.[11] Both were invited to attend enlarged meetings of the Central Committee's Political Bureau and they helped initiate and lead the most important political campaigns during the late years of the Cultural Revolution. They bypassed the regular party hierarchy and reported directly to radical leaders Jiang Qing and Zhang Chunqiao.

Tsinghua was important to radical leaders because it was located at the apex of the political and academic credentialing systems that had selected and trained the Red experts they deemed to be the most dangerous agents of capitalist restoration. Both credentialing systems were viewed with great suspicion. With Mao's support, the radical faction turned Tsinghua into a laboratory for implementing new policies designed to undermine the political and cultural foundations that underpinned the elite status of both party officials and intellectuals. The radicals used the university as a base from which to extend their power and promote their policies to places where they had far less influence. The strength of the radical faction at Tsinghua makes the university atypical, but also makes it an especially informative case. Because the radicals stridently championed Cultural Revolution slogans about eliminating elitist education and curbing the bureaucratic power of party officials, it is of particular interest to examine how they ran an educational institution in which they held sway.

This chapter will examine the peculiar system of governance established at Tsinghua during this period (Chapters 7 and 8 will examine education and admissions policies, respectively). Power at the university was divided between the workers' propaganda team, composed of workers and soldiers drawn from outside the school, and veteran university officials. The propaganda team was charged with mobilizing students and workers to criticize their teachers, supervisors, and university officials. The result was a tumultuous system of governance very much at odds with the conventional practice of ruling Communist parties, including the CCP before the Cultural Revolution, which had been guided by ideals of monolithic unity and a clear hierarchy of authority. I will examine how this system functioned in practice and suggest reasons why it continued to reproduce familiar problems of political dependence. I will then consider how the Tsinghua system fit into wider patterns of governance around the country, which I suggest fostered a division of power between administrators and rebels.

Supervising Cadres and Teachers from Above and Below

The first tasks assigned to the workers' propaganda team that was sent to Tsinghua were to reestablish order, rein in the contending factions at the university, and create new leadership bodies. The team established its authority by harshly suppressing all potential opposition, and its authoritarian style was reinforced by team members' understanding that they were charged with occupying and transforming an institution that had been under the domination of "bourgeois intellectuals." The two contending student-led factions were dismantled, efforts to sustain factional activity were suppressed, and many factional leaders and activists were investigated and punished.[12] After a period of political study, most students were sent to military-run farms for a period and then assigned permanent jobs, typically in industrial enterprises. Teachers and administrators were subjected to a wider effort to gain quiescence—the campaign to "Clean-up Class Ranks" (*qingli jieji duiwu*). The workers' propaganda team reexamined the personal histories of faculty members and university cadres from before the 1949 Revolution through the Cultural Revolution. Of some six thousand Tsinghua employees, 1,228 were investigated and seventy-eight were declared to be "class enemies." According to a semiofficial school history, during a two-month period at the height of the campaign, ten people died as a result of persecution.[13] Most of Tsinghua's cadres, teachers, and staff were then sent to work in a rural "May 7 cadre school" farm in Jiangxi Province, returning only after a new cohort of students arrived in 1970 (see Chapter 7).

REDS SUPERVISING RED EXPERTS

By sending in workers' propaganda teams to supervise university cadres, Mao reproduced the Red-over-expert power structure that had characterized the first years of Communist power, when poorly educated Communist cadres supervised incumbent White administrators and experts. Like the party cadres who took charge of Tsinghua in 1952, the workers' propaganda team arrived at the university in 1968 with a revolutionary mandate to transform an institution considered to be under bourgeois control. Only this time, those being supervised and transformed were primarily Red and expert cadres who had been cultivated during the first seventeen years of Communist power. Most of the military officers and factory workers who made up the workers' propaganda team did not have much education, but this was precisely why, in Mao's estimation, they had the credentials necessary to transform universities. Because they had little formal education, they were not bound by the interests and prejudices that might incline intellectuals to maintain the status quo in educational institutions. The workers and soldiers assigned to participate in the propaganda team were charged with "occupying and transforming the superstructure," that is, universities that had remained under the control of bourgeois intellectuals.

Initially, the propaganda team at Tsinghua was a huge group, composed of 5,147 factory workers and 105 military officers, but its size was substantially reduced after a few months.[14] While most of its top leaders—like Chi Qun and Xie Jingyi—were soldiers, industrial workers also played key roles, and the vast majority of team members were ordinary workers from local factories. The workers' propaganda team was Red but not expert. Most were party members and all had been active in party-led organizations in their factories, such as the Youth League, the trade union, the women's federation, and the militia, but the best educated only had a middle school education, and many had only been to primary school. Although the propaganda team was to remain at Tsinghua permanently, its membership changed. With the exception of the very top leaders, members of the team rotated in and out of the university, returning to the factories or military units from which they had come, and while they were at the university they continued to receive their wages from their original units. They were at Tsinghua to supervise university administrators, not to become administrators themselves.

The workers' propaganda team rehabilitated most university cadres and many were appointed to positions similar to those they held before the Cultural Revolution.[15] In January 1969, the leaders of the propaganda team established a revolutionary committee to administer the university, and a year later they reestablished the university party committee, which had ceased to function in 1966. Liu Bing was reappointed deputy party secretary and was charged with handling the party's organizational affairs. He Dongchang, who before the Cultural Revolution had been in charge of academic affairs, was again given this role, although his formal title was deputy director of the university "teaching reform committee," the top position being reserved for a representative of the propaganda team.[16] By 1972, the standing committee of the school party committee was made up of nine members of the propaganda team and eight Tsinghua veterans. Chi and Xie, the leaders of the propaganda team, were secretary and deputy secretary, respectively.[17]

While the leaders of the workers' propaganda team were closely associated with the top radical leaders in the CCP, veteran university officials—including Liu Bing and He Dongchang—were closely associated with more conservative party leaders at the center. When the radical faction was ascendant nationally, so was the power of the workers' propaganda team at Tsinghua, and when the conservatives were ascendant nationally, the power of the veteran university cadres also rose. Neither the propaganda team nor the veteran university cadres, however, were politically homogeneous—some veteran cadres aligned themselves closely with the propaganda team leadership, while a number of propaganda team leaders ended up siding with the veteran university cadres. The Tsinghua cadres and the members of the propaganda team each derived their power from different kinds of qualifications. On one side, the university cadres had both strong cultural and political qualifications; they were among the best-educated people in the country and many ranked higher in the party hierarchy than did the propaganda team interlopers. Moreover, Tsinghua was their territory and they had built the university party organization that the propaganda team leaders now—tenuously—commanded. On the other side, the power of the propaganda team was based solely on political qualifications, as its members had little education. Moreover, their political qualifications were of a distinctly Cultural Revolution variety. Even the top leaders of the team did not originally rank high in the party hierarchy; their political authority stemmed not from their party position, but rather from their assignment by Mao to run the university. At the time, however, this gave them greater authority than higher-ranking university cadres.

Mai Qingwen, a high-level veteran university cadre, recalled that during this period Mao's personal authority was much greater than that of the party organization. Rebel attacks had undermined the party's *weixin*, a term that can be translated as prestige, popular trust, or authority. "During the early period of the Cultural Revolution the *weixin* of the Communist Party and the Communist Youth League fell sharply," Mai explained. "All the leading cadres were criticized, and whether or not the criticisms were correct, the conclusion was that they were all bad." Because the propaganda team had been sent by Mao, Mai said, its authority outstripped that of the university party organization. "The workers' propaganda team used the party organization, but the party itself had no *weixin*. The workers' propaganda team led the party, not the other way around—Chi Qun was first of all the head of the workers' propaganda team, and then the party committee secretary."[18]

Cheng Yuhuai, another high-level Tsinghua official, recalled how the Cultural Revolution upset the hierarchy of ranks within the party. "He Dongchang was a ministry-level cadre," he explained. "Under the ministry level, there were bureau, department, and division levels. Chi Qun was only a division-level cadre, way below He Dongchang. . . . It was really a mess (*daluan*)—the higher your rank, the more likely you were to be overthrown, and low-rank people got promoted. . . . It was complete disorder."[19] Although the university party organization was rebuilt, it was hardly the formidable and monolithic political machine it had been before the Cultural Revolution. The orderly mechanisms of decision making, the normal chain of command, and the conventional systems of promotion, seniority, and rank were all disrupted. Moreover, while the leaders of the workers' propaganda team took control of the Tsinghua party organization, they did not trust it, and they created a network of power that largely bypassed it. "Chi Qun maintained direct relations with [propaganda team representatives]," recalled Cheng. "They would report through a different system. They had the power—if they disagreed, it couldn't be done."[20]

The rehabilitated university cadres retained much of the responsibility for running day-to-day affairs at the university, including teaching, research, and factory production, but ultimate decision-making power was in the hands of the workers' propaganda team. "The military representatives didn't understand education, so we took care of it, and then they criticized us," said Tong Yukun, another high-ranking veteran Tsinghua official.[21] Hong Chengqian, who served as a manager at the university's machinery factory, complained, "I was a cadre who had been rehabilitated, but I was not really a leader. I just helped the workers' propaganda team leaders—I could make suggestions, but . . . they wouldn't let any Tsinghua people have any leadership power."[22] Today Tsinghua faculty members often tell two types of stories about representatives of the workers' propaganda team. Some team members, they say, recognized that university cadres and teachers knew more about education than they did, and they were reasonable and cooperative. Others, despite their ignorance, insisted on imposing their will. Fang Xueying, the daughter of a revolutionary cadre who studied at Tsinghua during the early 1970s and now teaches at the university, described the second type. "The leader of the workers' propaganda team in our department was very fierce—he had a bad temper and he liked to bang his fist on the table and yell at people," she told me. "He was a country hick (*tubaozi*), he didn't have any education. . . . [But] he thought he was really something—Chairman Mao had sent him to Tsinghua."[23]

Zhuang Dingqian, a middle-level party cadre whose parents were both university professors, told me that he agreed with the principle of working-class leadership, but in his view this meant that "the working class should lead through the Communist Party." It did not mean that actual workers should lead a university. "During the Cultural Revolution . . . we all had to listen to the workers' propaganda team. I think it was ridiculous—what did they know? . . . Mental labor and manual labor don't understand each other. If you let workers, that kind of workers, lead the school—it was a joke."[24] Tsinghua simmered with tension during the Cultural Revolution decade as university cadres and teachers resented the heavy-handed leadership and radical policies of the workers' propaganda team. Nevertheless, most cadres and teachers worked hard to implement Cultural Revolution policies and accommodate the demands of the workers' propaganda team.

MASS SUPERVISION

Before the Cultural Revolution, the Tsinghua party organization was unified and there was a very clear hierarchy of authority. At the bottom, students were expected to submit to the authority of their teachers, and workers in university-run factories were expected to submit to the authority of their supervisors. After the arrival of the workers' propaganda team, this one-way hierarchy of authority was replaced with a system in which the propaganda team supervised university administrators and teachers from *above*, while mobilizing students and workers to criticize them from *below*. The simplified diagram presented in Figure 6.1 compares the lines of supervision before the Cultural Revolution with those during the late Cultural Revolution years.

Under the workers' propaganda team, leadership committees at all levels included representatives of the propaganda team, veteran university cadres, and the "masses" (rank-and-file university workers, teachers, and students). At the top, the university revolutionary committee was composed of propaganda team leaders and top university officials, along with a number of ordinary workers and teachers, and the same "three-in-one" (*san jiehe*) principle was followed at the department level. The policy of including workers, teachers, and students in leadership bodies was described as "mixing in sand" (*chan shazi*), a reference to the practice of adding sand to cement. On the one hand, this metaphor referred to integrating workers and peasants into bodies composed of intellectuals, and the contrast was between those who did mental labor and those who did manual labor. Under the same rubric, however, rank-and-file

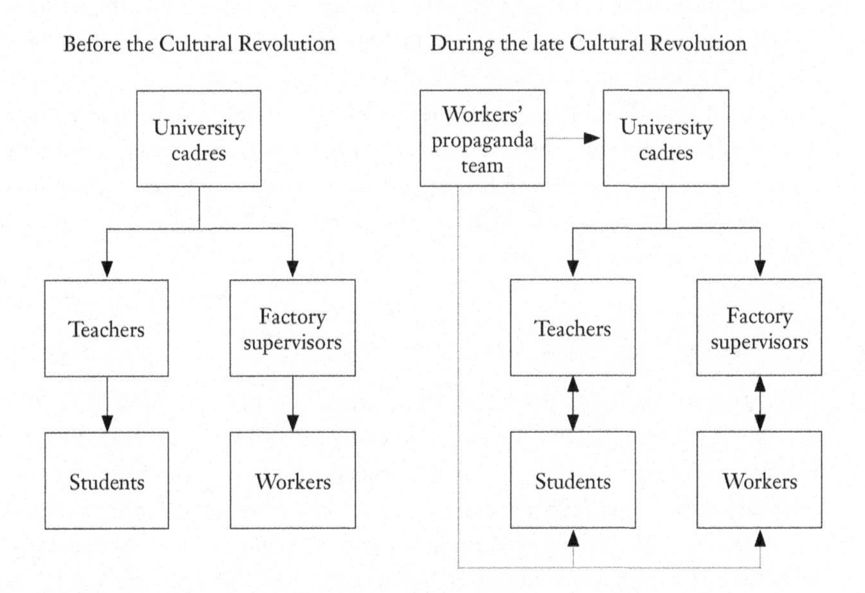

Figure 6.1 Supervision regimes at Tsinghua University

teachers were promoted to serve as members of top university leadership bodies, as "representatives of the masses," and in this case, the idea was to introduce nonleaders into leading bodies.

I asked Lin Jitang, a former student factional leader who had been retained to work at the university and was subsequently promoted by the workers' propaganda team to a top leadership position, whether the "three-in-one" policy was simply tokenism. Did campus workers really have any influence in university leadership bodies? "At that time," he recalled, "there were Tsinghua workers on both the [university] party committee and the [university] revolutionary committee. They listened to their opinions; the workers' propaganda team thought very highly of workers. At that time, the status of workers was very high."[25] As I spoke with people who studied and worked at Tsinghua during the late years of the Cultural Revolution, I began to appreciate that under the peculiar power dynamics of this period, members of the workers' propaganda team were involved in a daily battle with veteran Tsinghua cadres and teachers, and in this battle they mobilized—under the rubric of "mass supervision"—the university's workers and students as allies.

Although "mass supervision" over cadres had long been a Communist slogan, it was usually only invoked during political movements against corruption and bureaucracy. During the seventeen years that preceded the Cultural Revolution, mass supervision found little practical application in Tsinghua's everyday political and administrative machinery. Then, during the early years of the Cultural Revolution, as we have seen, it became a rallying cry of the radical faction, as they subjected university cadres to mass "struggle and criticism" meetings. Now, during the late years of the Cultural Revolution, mass supervision became part of the routine methods used by the workers' propaganda team to govern the university.

STUDENTS SUPERVISE TEACHERS

From 1970 to 1976, classrooms at Tsinghua and other universities were filled with students who had been recommended by factories, rural communes, and military units (see Chapter 8). The workers' propaganda team at Tsinghua considered these "worker-peasant-soldier students" (*gongnongbing xueyuan*) to be more politically reliable than university cadres and faculty members, who were largely from prerevolutionary elite families and had been trained by the old Tsinghua. The new students were, according to a contemporary slogan, not only supposed to attend the university, but also to administer and reform it. At a mass meeting held to welcome the first cohort of worker-peasant-soldier students to Tsinghua in 1970, Zhu Youxian remembered that leaders of the propaganda team told him and the other new students: "You come from the ranks of the workers, peasants and soldiers. Tsinghua is a dominion controlled by intellectuals (*zhishifenzi yitong tianxia*), so you have to get involved in politics."[26]

The basic organization of the university was changed so that students could participate in the management of their departments. Before the Cultural Revolution, as we have seen, both students and teachers had been highly organized, but separately. While students managed their own affairs (study, recreation, welfare, and so on), faculty members handled teaching and academic affairs. Departments were run by faculty committees and majors within departments were managed by faculty "teaching and research groups." During the Cultural Revolution, in contrast, students and teachers were integrated into the same organizations, and students took direct part in managing teaching and research. Student classes were each composed of approximately twenty-five students who lived and studied together for the entire time they were at the university, as they had before the Cultural Revolution, but now a group of teachers was permanently assigned to each class. These classes became the basic organizational units for both students and faculty members. Classes in the same cohort and major comprised a "small teaching group," which included students, teachers, and a representative of the workers' propaganda team. Each major, composed of three small teaching groups (one for each cohort), was run by a "teaching reform group" made up of representatives of the faculty, students, and the propaganda team. Departments, made up of several majors, were run by revolutionary committees, which also included representatives of the faculty, students, and the propaganda team. The organizational integration of teachers and students was further cemented by the fact that nearly half the students were already party members, and students and teachers typically belonged to the same party branches. In fact, some party branches were led by students.

This cellular structure certainly limited the variety of instructors and courses available to students, but it facilitated student participation in decision making. Students participated in designing curricula and solving teaching problems, and the small teaching groups met regularly to discuss academic, organizational, and political affairs. According to Fang Xueying, she and other students were hardly passive participants. "The teachers would report and we would comment; sometimes we rejected the entire report, and sometimes we would agree with some parts and reject other parts," she reported. "The students' enthusiasm for education reform was very high. We would discuss the direction of our major—ultimately, what kind of people should we be training?"[27]

The content of these discussions was highly circumscribed by the ideological and political orientations established at the top levels of the CCP. "Although there were debates at the meetings and big character posters expressed different ideas, generally they followed the fundamental points set by the center," Wei Xuecheng, a veteran teacher, told me. Nevertheless, the debates were far from meaningless. They involved education policy issues that were at the center of the radical-conservative factional conflict, including how to implement "open door" teaching methods, how much emphasis to place on theoretical as opposed to practical curricula, how to evaluate student learning, whether or not to divide students into fast and slow classes, and so on. As a result, both students and teachers found themselves on the front line of debates that defined Chinese politics during this period. All of this political activity, Fang Xueying said, trained worker-peasant-soldier students how to write, speak, and organize. "We were supposed to attend, administer, and reform (*shang guan gai*), to criticize the teachers and the old education system. Every evening we had political meetings, we discussed issues, and we wrote big character posters."[28]

As the influence of the radical and conservative camps ebbed and flowed and central policy guidelines swung from Left to Right and back again, some individuals attempted to defend deeply held beliefs, while others shifted nimbly with the prevailing winds. Some students inevitably pushed for the most radical interpretation of education policies, often with the support of the workers' propaganda team representative, while many teachers found themselves in the position of arguing to preserve aspects of conventional teaching practices. Teachers were handicapped in the debates, especially if they called for moderation because they could easily be accused of "wearing new shoes, but walking on the old path." Luo Xiancheng, a peasant youth who began studying at Tsinghua in 1970, recalled: "[When] we had meetings to debate education problems, the students and the workers' propaganda team representative spoke openly, but the teachers didn't—they were the objects of reform. It was not an equal discussion," he admitted. "The teachers were more cautious—it was not that they didn't dare talk, but they were more careful."[29] While students in the past had often been intimidated in the presence of teachers, the situation was now reversed. Wei Jialing, a veteran teacher, described the changes as a harmful inversion of conventional teacher-student relations. "Before the Cultural Revolution, teacher-student relations were traditional Chinese teach-and-be-taught relations, so relations were pretty good—students respected their teachers and teachers cared for their students," she told me. "During the Cultural Revolution, things changed—teachers became the objects of supervision and students became the masters of the school."[30]

Under the workers' propaganda team, criticism of teachers and university officials from below became routine, rather than extraordinary, as it had been in the past. The difference was noted by Yang Yutian, a veteran teacher: "At that time, [the workers' propaganda team] asked students to criticize bourgeois intellectuals, but it was not like the [1957] Anti-Rightist campaign. Then, they called some people Rightists and treated them like enemies," he recalled. "But during the worker-peasant-soldier student period . . . there were no enemies. They would criticize intellectuals in general; it made you uncomfortable, it made you careful when you spoke, but they didn't put a hat on you. We got used to being criticized."[31] Yang's testimony should not be interpreted as an indication that political repression in the 1970s was mild; it certainly was not. Rather, Yang was indicating that for teachers and cadres, criticism from below had become a mundane, everyday activity. In the past, in contrast, criticism from below had usually been limited to campaigns, it had been directed against a relatively small number of individuals, and it had serious consequences.

Traditionally, teachers were expected to conduct themselves in a severe and strict fashion so as to evoke the proper deference from students, and laughing or joking with students was seen as undermining a teacher's dignity and authority. During the Cultural Revolution, these ideas were criticized as the "teacher's code of dignity" (*shidao zunyan*), a term that mockingly referred to Confucian codes of conduct for men in positions of authority. Some students relished the opportunity to diminish the social position of senior professors. Long Jiancheng, a foreign student in the water conservation department, remembered the daily trials faced by Zhang Guangdou, a Cornell-trained hydraulic engineering professor who had led the team that designed the huge Miyun Reservoir in 1958. "Students used to treat him badly," Long recalled. "A female student would hit him all the time—she wouldn't hurt him, but she'd just go up and hit him." Before the Cultural Revolution, Zhang had been one of Tsinghua's most powerful and respected professors; now he had to maintain good humor while fending off student taunting. "They made fun of him, and he made fun of them," Long said, adding, "[but] he wasn't beaten down, and the students respected him."[32]

Despite the harsh inversion of authority between students and teachers, a number of students and teachers reported that they developed very close relations during this period. Fang Xueying, who experienced Tsinghua as a student during the early 1970s and as a teacher since then, compared student-teacher relations today with those when she was a student. "Then the teachers' status was not high like it is today—if anything, the students' status was higher," she told me. "The teachers agreed they had been trained under the old system and needed to be reformed, so they were willing to be together with the students. They didn't have as much authority as teachers have today, but the relationship between teachers and students was closer."[33] Liang Yousheng, who studied at Tsinghua in the 1950s and began teaching at the university after he graduated, made a similar observation. "Before the Cultural Revolution . . . students were very respectful of teachers; the hierarchy was very clear—it was a relationship between lower and higher. [During the Cultural Revolution,] most of the worker-peasant-soldier students were very grateful to the teachers, but it was not a hierarchical relationship because we lived, ate, and worked together," he recalled. "Their feelings toward the teachers were very close and strong. If the teachers made a lot of effort to teach them, they really appreciated it."[34] Fang, Liang, and others noted that students and teachers became closer especially during the periods they lived and worked together in factories. "When we were at the factory, I lived together with my students, eight to a room," Liang told me. "We shared everything; we could talk about anything."[35]

WORKERS SUPERVISE FACTORY LEADERS

The management system at Tsinghua's campus factories underwent similar changes. Before 1966, factory workers—like students—had been highly organized and small groups of workers largely managed shop-floor production, but beyond these small groups their role was very limited. This changed dramatically during the early years of the Cultural Revolution. While student-led factions were contending for control of the university, workers elected temporary committees to run the campus factories. Chen Jinshui, a veteran worker, described what happened at Tsinghua's equipment factory. "All the leaders in our work unit were overthrown, including the factory manager, the deputy manager, the party secretary, and the deputy party secretary. We elected a new leadership group . . . four were workers and one was a technical employee," he told me. "The group changed—if someone was no good, someone else was elected." At the time, Tsinghua's factories were also split into contending factions, but voting did not simply follow factional lines. "Factions were important, but they were not the most important thing," Chen recalled. "The most important thing was whether or not you trusted the person—whether they were honest, had the ability to get things done, and had good relations with the masses."[36]

In the summer of 1968, when the factional conflict at Tsinghua turned violent, most workers abandoned the campus factories. After the workers' propaganda team took over the campus in August, it organized new revolutionary committees to run the factories, headed by representatives of the propaganda team. Veteran managers were typically brought back to assume leadership positions, but workers, including some of those who had been elected to the temporary leadership bodies, were also appointed to serve on the revolutionary committees. Even after the old managers were reappointed, workers retained substantial influence in factory affairs, and relations between managers and workers remained an issue of mass discussion. "[We criticized the managers for] being divorced from practice, for subjectivism," Chen recalled. "We criticized bureaucratism—if you don't understand something, you should ask those below you, otherwise it won't work."[37]

Hong Chengqian, a factory manager who was overthrown at the beginning of the Cultural Revolution, recalled the period with ambivalence. "Before the Cultural Revolution, the workers were supervised by others (*bei guanli*); then, at the beginning of the Cultural Revolution, the workers became the masters." After the workers' propaganda team arrived at Tsinghua, Hong became a manager again, but the circumstances had changed. "In those days, the workers rebelled against the rules—there was a tendency toward absolute egalitarianism, toward anarchism," he recounted. "The workers would say, 'Why is it that when you become a cadre, you don't work anymore?' With that kind of thinking, it got very chaotic."

At first, mass meetings were called to make many decisions, but that method, Hong argued, was not practical. "One hundred people can't lead," he told me, offering an example he said he often used to explain the problem to workers. "Right after the workers' propaganda team came, they called a big meeting to discuss working hours—what time to start and quit work and break for lunch. The entire workforce met and discussed the question all day, but couldn't come to any decision. The problem was that some lived at school, some lived in the city, some were single, some had children—so they all had different ideas. In the end, they couldn't decide, so they just followed the school rules." Gradually, the number of big meetings decreased. "The meetings became more of a formality," Hong recalled. "To tell the truth, the workers weren't really that interested in attending." Despite the decline in enthusiasm for mass meetings, however, workers continued to play a role in management, and workers' representatives had to be consulted in production decisions. "We had to discuss everything with the workers. They were enthusiastic about discussing production. Of course, the purpose then was to criticize intellectuals, and that's what the workers did. The common expression was: 'It's easy for an intellectual to draw a line on paper, without considering the difficulties encountered by the workers in actually producing it.'"[38]

Institutionalized Rebellion

The workers' propaganda team played a role very similar to that of the party work teams sent to work units to deal with problems of corruption and cadre abuse of power during earlier political movements, only it became a permanent fixture at the university. As such, the propaganda team was able to develop a system of governance based on regular mechanisms of mass supervision over university cadres and teachers. But this model of mass supervision—as it was implemented at Tsinghua—had a critical flaw: the propaganda team itself was not subject to supervision from below. In principle, the team was not supposed to be above criticism by subordinates. Its leaders sharply criticized the previous university leadership on this score and they set high standards for themselves. "We must be modest and prudent, guard against arrogance and impetuosity, humbly learn from the masses, listen to the opinions of the masses, and accept the supervision of the masses," the leaders of the team wrote in an article published in the *People's Daily*. "Since we are in a leadership position, we can never stay aloof and put on bureaucratic airs. If you can only criticize others but are not able to dissect yourself, if you have ears only for flattering words but not for unpleasant ones, if you act as if you are the buttocks of the tiger that cannot be touched, these are all . . . manifestations of a bourgeois lifestyle and work style. . . . We must consciously put ourselves under the supervision of the broad masses, and the workers' propaganda team members who shoulder major leadership work should in particular take this course."[39]

Despite this rhetoric, however, in practice the workers' propaganda team was no more positively disposed to criticism from below than the Jiang administration had been. Few members of the Tsinghua community dared to criticize the leaders of the propaganda team in public, and those who did suffered consequences that dissuaded others. As a result, the culture of political dependence that had flourished at the university before 1966 reemerged during the tenure of the propaganda team, but in a different form. The compliant style of political activism that had been fostered by the party and the Youth League before the Cultural Revolution was now replaced by a new style of political activism that might be called sycophantic rebellion.

During the late years of the Cultural Revolution decade, the radical faction of the CCP paid tribute to the heroic image of rebel fighters during the early years of the Cultural Revolution, and aspiring activists were exhorted to emulate them by displaying their own "rebel spirit." At Tsinghua, however, the meaning of rebel spirit was distorted by the actual power relations at the university, as members of the propaganda team—who held ultimate power—mobilized students and workers to criticize members of the pre–Cultural Revolution establishment. Activists "rebelled" against old leaders to curry favor with new leaders. The defects of this type of sycophantic rebellion can be found in many of the radical faction's endeavors during this period, including in their efforts to reform the political credentialing system and to mobilize mass protests.

RADICAL ATTEMPTS TO REFORM THE POLITICAL CREDENTIALING SYSTEM

After the revival of the party organization in 1969, radical leaders, hoping to shore up their weak position in the party, pushed for massive recruitment of new members. At Tsinghua, for instance, the workers' propaganda team presided over the recruitment of 3,271 new members between 1969 and 1976, and the size of the university party organization more than doubled during this period.[40] For the radicals, however, the point was not simply to increase the size of the party, but to fill its ranks with new members more sympathetic to their aims. They therefore insisted on changing the recruitment criteria. Before the Cultural Revolution, those who applied to join the party were expected to demonstrate, in addition to public spirit and commitment to Communist doctrine, a willingness to comply with bureaucratic authority. In contrast, Cultural Revolution activists, like Kuai Dafu, established their rebel credentials by demonstrating exactly the opposite quality—their willingness to challenge bureaucratic authority. Now the radical faction hoped to import the insubordinate attitude of the rebel movement into the revived Communist Party, and they insisted that the main criterion in recruiting new members should be rebel spirit.

The height of the recruitment drive coincided with a radical political offensive that followed the Tenth Party Congress in August 1973. During this offensive, the radicals called for a renewal of the spirit of the Cultural Revolution, encouraging their followers to "go against the tide" (*fan chaoliu*), that is, to resist what they saw as a retreat from the principles of the Cultural Revolution. In the national press, the radicals encouraged party branches not to recruit "obedient tools" who obeyed their superiors without question, but rather to seek out young activists who displayed rebel spirit, who "thought independently" (*duli sixiang*), who had "horns on their heads and thorns on their bodies" (*toushang zhang jiao, shenshang zhang ci*), and who would not shy away from raising criticisms of the leadership. They stridently warned against the tendency of the party to become conservative, referring to this tendency with the pejorative terms *haoren dang* and *yewu dang*. *Haoren* in this context elicited the image of *laohao ren* (good guys), that is, uncontentious souls who value getting along over sticking to principles. *Yewu* referred to economic and technical work (as opposed to political work), and the term *yewu dang*, or "party of technical work," was employed to criticize tendencies to focus on managing society, rather than transforming it. The job of the Communist Party, in their view, was to lead the ongoing revolution, not to consolidate the existing order. Communists, therefore, were to be fearless revolutionaries, not compliant and careerist bureaucrats and technicians.

Not surprisingly, many local party leaders were not eager to recruit new members who had horns on their heads and thorns on their bodies. To urge them in this direction, the national press was filled with articles that recounted model experiences in which party branches finally recruited individuals who previously had been considered "troublemakers," "argumentative," and "unable to take leadership" after recognizing that these individuals possessed the kind of fighting spirit needed to combat revisionism. It is not difficult to imagine that in most party branches this was an uphill battle. Hong Chengqian, the Tsinghua factory manager, explained how party culture was adverse to people who raised criticisms. "During all periods, most of those who joined the party had good moral character (*pinzhi*)—they were honest and well behaved (*laoshi*) and they did good work," he told me. "But during the Cultural Revolution, a few of those recruited liked to criticize people. We called them 'verbal revolutionaries' (*koutou geming pai*)—they could talk well, but they did not do good work." In Hong's estimation, good party members worked hard and were well behaved, while rebels did not work hard and liked to criticize others. In this duality, there was little room for rebels to be good party members. It was clearly difficult to try to impose rebel political criteria that grew out of the period of freewheeling factional contention on a party steeped in a culture of hierarchically organized discipline. "Rebels (*zaofan pai*) still had difficulty getting into the party," Hong added. "They had to be voted in—the party [branch] members had to agree. Most party members had joined before the Cultural Revolution and they looked down on rebels."[41]

The meaning of rebel spirit, however, had also changed. This can be seen in the way one student remembered how the criterion rebel spirit was considered when membership applications were discussed in her party branch. "Formally, the [recruitment] process was the same as it is today, but the standards were different," recalled Fang Xueying. "During the Cultural Revolution they looked at whether or not you had rebel spirit. . . . [That is,] whether or not you criticized the teachers and cared about politics—whether or not you were active in criticizing people." I asked Fang whether criticizing the leaders of the workers' propaganda team would have been seen as an indication of the appropriate spirit. "Actually, the leaders always said they wanted people with rebel spirit," she replied, "but in fact they wanted people who would listen to them and criticize others—criticize the old leaders. Whoever becomes a leader likes those who are obedient (*tinghua*)."[42] Thus, displaying rebel spirit at Tsinghua during this period meant following the lead of the workers' propaganda team, a spirit that had little in common with the insubordinate attitude celebrated during the early years of the Cultural Revolution.

The sycophantic rebel behavior fostered by the workers' propaganda team did not prove very popular, at least not among party members in Fang Xueying's class. She recalled how members of her party branch rejected the application of one student who was seen as too eager to criticize others in order to please the soldier who represented the propaganda team in their department. "Actually, it was also very important that you had good relations with the other students. . . . For instance, one student had rebel spirit, he was very active in criticizing, but we refused to let him in. He had very good relations with the military representative and the military representative promoted his application. He told us to admit him, but the students didn't like him, so they refused to vote him in. . . . We thought he wanted to curry favor (*pai mapi*), that he would do anything to please the leadership."[43] This combination of servility and rebellion could alienate many people for different reasons. Many students and workers did not like people who catered too much to those in positions of power. At the same time, the party's culture remained much more amenable to uncontentious and career-minded *laohao ren* than to rebels of any kind. It would be wrong to say that everyone who raised criticisms of people and phenomena identified with the pre–Cultural Revolution order at Tsinghua was simply currying favor with the workers' propaganda team; many people were sincerely committed to advancing the Cultural Revolution agenda. Nevertheless, the system of supervision from above and below created conditions in which all criticism of this type could be interpreted as currying favor, and it was no doubt difficult even for those raising criticisms to sort out their own motivations.

LACKLUSTER POLITICAL CAMPAIGNS

The radical faction sought every opportunity to launch new movements against the party officialdom, and they won Mao's support for major mass campaigns in 1974 and 1976. Both campaigns were initiated at Tsinghua. In late 1973, Chi and Xie mobilized students and teachers to criticize He Dongchang for promoting conventional teaching practices at the university, a campaign that paved the way for a major nationwide movement to Criticize Lin Biao and Confucius in 1974. Then, in the fall of 1975, a Great Debate on the Education Revolution was initiated at Tsinghua, with Liu Bing as its main target, and it was converted the following year into a nationwide campaign to criticize Deng Xiaoping and oppose the Right Deviationist Wind to Overturn the Correct Verdicts of the Cultural Revolution. In both cases, education policies were the initial focus of campaigns that were later extended to include all of the major issues that divided the party's radical and conservative factions.

During the anti-Deng campaign, the national press—under radical control—featured increasingly inflammatory rhetoric, denouncing party officials in language that was even stronger than that used in the early years of the Cultural Revolution. At that time, unofficial rebel newspapers had described their enemy as a "privileged stratum"; now the country's leading newspapers were calling on people to do battle with an entrenched "bureaucratic class." The methods of mobilization used, however, bore little resemblance to the freewheeling factional contention of the early years of the Cultural Revolution. Instead, the radicals largely relied on the kind of top-down bureaucratic methods of mobilization that Communist parties in power had conventionally employed. In 1966, Mao had encouraged students, workers, and peasants to form their own "fighting groups" and to "link-up" with other fighting groups; in 1976, in contrast, students and workers were instructed that the struggle was to be conducted under party leadership. Early in the 1975–76 campaign, when the focus was still on education issues, the party committee in Tsinghua's civil engineering department described how the campaign was organized. "The current big debate on the education revolution has been conducted step by step, under the leadership and according to the unified arrangements of the party committee," department leaders wrote. "We do not have 'link-up' inside and outside the department and inside and outside the university. We have no 'fighting groups.' We build no mountain strongholds, and we do not stop work or stop classes. 'All actions are taken as they are ordered.' This has ensured the smooth and healthy development of the movement."[44]

The contrast between the ends and the means of the campaign could not have been more dramatic: people were called upon to rise up against the bureaucratic domination of party officials by participating in a movement under the direction of the local party committee. Although the movement was a manifestation of a momentous political confrontation between the radical and conservative wings of the party and contemporary press reports described it as filled with "vitality and vigor," students who experienced the campaign at Tsinghua remembered it as considerably less lively. Zuo Chunshan emphasized the compulsory nature of participation. "We were ordered from above to attend meetings," he recalled. "You had to participate actively, otherwise you'd be accused of only being interested in studying your vocation, so you had to speak up actively in meetings and you had to follow the line." Long Jiancheng, a foreign student studying at Tsinghua in 1976, remembered the political meetings that accompanied the movement as boring. "They were like

Bible classes—we would start by singing a song, then we would read Engels or Lenin and discuss how to apply their ideas, then we would sing another song, and that was the end," he recounted. "There was not much enthusiasm." Even Chi Qun and Xie Jingyi, speaking to a study meeting in Tsinghua's electrical engineering department in the spring of 1976, admitted that morale in some units was low and that people "appear war-weary and sullen."[45]

Although the 1975–76 campaign was in many ways a continuation of the assault on the party officialdom that Mao had launched in 1966, it lacked the rebellious energy of the early years of the Cultural Revolution, when the party organization was paralyzed, and students, teachers, and workers formed their own fighting groups. Then, tens of thousands of people poured onto the Tsinghua campus to read the latest big character posters, and they crowded into meeting halls, hanging on every word of the contentious debates. In 1976, thousands of people also came to Tsinghua to read big character posters, but this time they were trucked to the university by their work units, and the content of the posters was orchestrated by the workers' propaganda team. The campaign not only lacked the unruly energy of the early years of the Cultural Revolution, but it had little support from the party apparatus, which was largely in the hands of unsympathetic officials. The result was a feeble campaign doomed to failure.

Institutionalized Factional Contention

In Mao's waning years, members of both the radical and conservative factions sought to win his favor, and he first supported one faction and then the other. This ambivalence has long puzzled scholars. As I investigated the contentious politics of these years, I began to discern a pattern that made sense of this ambivalence. At every level, it appeared, factional contention was being institutionalized. Overthrown party officials were rehabilitated, but members of the rebel groups that had proliferated during the early years of the Cultural Revolution were also given positions of power. Moreover, the pattern of appointments indicated that these two types of individuals were charged with distinct tasks: veteran cadres were put in charge of running the political and economic bureaucracies that allowed the country to function, while rebels were given institutional means to mobilize political campaigns against these officials, pressing Mao's radical agenda. It seemed that Mao was deliberately constructing a system of governance that pitted *rebels* against *administrators*. This scenario places the factional rivalry that dominated political life in China during the late Cultural Revolution years, which has conventionally been treated as a struggle for succession, in a different light. Perhaps this contention was not simply a means to an end, but an end in itself.

I observed this type of political setup at close range at Tsinghua. The entire educational system, however, also seemed to be organized in a similar fashion, and indications of this kind of logic were plentiful outside of the realm of education as well. In 1970, the State Council (China's top administrative body) convened a Science and Education Group to revamp education policy. Chi Qun was named deputy leader of the group and, with the backing of Deputy Premier Zhang Chunqiao (who was charged with overseeing educational affairs for the State Council), he became its most influential member. Chi steered the group toward advocating radical policies that eschewed elite training and instead focused on practical education for workers and peasants. The Science and Education Group, however, did not have administrative authority over the school bureaucracy; it could only exhort school officials to implement its initiatives. When the Ministry of Education was finally reestablished in late 1974, replacing the ad hoc Science and Education Group with a formal administrative bureaucracy, Mao rejected the entreaties of Jiang Qing and Zhang Chunqiao to appoint Chi as education minister and instead supported Zhou Enlai's more conservative nominee, Zhou Rongxin.[46] The latter energetically promoted more conventional education policies through the ministry and the formal education bureaucracy, while Chi Qun and Xie Jingyi, with the support of their radical patrons at the center, promoted ever more radical policies, using Tsinghua and other schools under their control as models.

The same type of division—between administrators and institutionalized rebels—could be seen at the provincial, local, and school levels. At each level, the administrative bureaucracies were filled largely with rehabilitated veteran cadres, but radicals were appointed to serve as "mass representatives" on the new revolutionary committees that nominally supervised the work of the administrators. As party committees were restored at every level, a number of radicals were also selected to join their ranks, and in 1973 many local radicals were nominated as delegates to the Tenth Congress of the CCP. Radicals, however, were seldom given regular administrative duties, and many were not assigned any duties at all.[47] When radicals were assigned specific duties, they were typically asked to reorganize unions and other mass organizations and carry out campaigns of study and criticism.[48]

Within schools, power was further divided by the introduction of worker-peasant-soldier propaganda teams. These teams usually had far less authority than Chi and Xie enjoyed at Tsinghua, and in village primary schools the corresponding positions were typically assigned to a single peasant representative, who in practice was often relegated to the functions of ombudsman and community liaison.[49] Nevertheless, the nominal duties of propaganda team members included supervision of school officials, and while these officials were ordered to carry out the more conventional policies handed down through the formal education bureaucracy, the radicals attempted to mobilize the propaganda teams to back their unconventional initiatives, with varying degrees of success.

During this period, education policy swung first Right and then Left in cycles. Mao insisted that the education bureaucracy be administered by conservative officials and he supported their efforts to "regularize" education, but he also supported a series of radical political campaigns against education officials and their policies.[50] As a result, conservative power was firmly based in the administrative bureaucracy, while radical power depended largely on Mao's personal authority. Organizational ties among the radicals were mainly informal, existing parallel to the regular party bureaucracy and often in violation of its rules. Thus, although the radicals were ensconced near the centers of power, their authority—which was considerable—relied on Mao's support and the understanding that they were torchbearers for his charismatic mission. Their exhortations were broadcast widely and bore the seal of the Great Helmsman, but implementation of their initiatives depended largely on the success of political campaigns because they had little administrative power.

This sketch of the institutional arrangements in the education field suggests that the type of factional contention I observed at Tsinghua was one manifestation of a wider pattern. Several scholars have suggested that the struggle between radicals and conservatives during this period was deliberately fomented by Mao to maintain a balance of power between competing factions, leaving him as the ultimate arbitrator.[51] The interpretation advanced in this chapter contains two contentions that build on this type of explanation. First, Mao was creating a peculiar system of governance based on institutionalized factional contention, and, second, this contention was predicated on a functional division of labor that pit rebels against administrators.

The Limits of Institutionalized Rebellion

The introduction of the workers' propaganda team into Tsinghua University created a system of divided power, in which the propaganda team, aligned with the radical camp in the center, was authorized to supervise veteran university cadres who were largely sympathetic with the conservative camp. This created the conditions for organizing supervision of cadres from above and below, which the propaganda team developed into an elaborate system of governance. In terms of the antibureaucratic goals of the Cultural Revolution, this model was not without accomplishments. It prevented the reestablishment of the conventional bureaucratic lines of authority, in which university officials, teachers, and supervisors had exercised unilateral supervision over students and workers. Now, they found themselves in relations of mutual supervision with students and workers, and the latter were involved in decision making to an unprecedented degree. It was not only possible for subordinates to criticize their superiors, they were expected to. The combined effects of this kind of mass supervision, the organizational integration of cadres and teachers with students and workers, and more egalitarian distribution (see Chapter 7) significantly reduced the gap between cadres and noncadres, and altered the distribution of power between the two groups.

The fact that no one could criticize the leaders of the workers' propaganda team, however, fundamentally truncated the significance of mass supervision. As a result, the main target of supervision was cultural power, rather than political power. The veteran cadres subjected to mass supervision were responsible for organizing teaching, research, and production, and they were criticized mainly for promoting intellectual elitism and favoring technical competence over political activism. The language for criticizing bureaucratic concentration of power was available (tirades against the bureaucratic ways of party officials were a staple of the national press), but this kind of criticism was not often heard in the university context. Moreover, when it was raised in the university newspaper, it rang hollow because real political power was concentrated in the hands of the propaganda team, which was largely exempt from criticism. While the radicals were developing an increasingly lucid theoretical critique of the concentration of power in the hands of Communist cadres, in practice they remained inextricably bound up with the political culture they were criticizing. At least this was the case at Tsinghua, where the radicals were in charge. Because in practice leaders of the propaganda team were exempted from criticism from below, mass supervision at the university ended up reinforcing—in a distorted form—long-standing patterns of political tutelage and clientelism. Efforts to transform the university under the heavy-handed leadership of the propaganda team ended up producing sycophantic rebel activism and lackluster political campaigns.

Tsinghua was far from typical. The university was located close to the center of national power, both geographically and politically, and the radical-conservative rivalry at the center shaped university politics more directly than it did in more remote places. Moreover, the prominence of Chi and Xie in national politics gave them tremendous power vis-à-vis veteran university officials. Propaganda teams at other schools lacked this kind of power (and they also were not necessarily as committed to the radical agenda). For these reasons, the power dynamics at Tsinghua were different from those in other work units. Studies about other schools, factories, and municipalities during this period have presented highly diverse accounts that have included strikes, protests, big character poster campaigns against party leaders, violent skirmishes, arrests, and local political coups and countercoups.[52] In many work units, radicals were largely relegated to an opposition role and they mobilized criticism of administrative cadres from *below*, a much more hazardous proposition than the kind of sycophantic rebellion that took place at Tsinghua. The evidence presented in this chapter, together with that presented in the handful of other investigations into local conflicts during this period, suggests the following observation: where rebel opposition was organized from below, it was weak and precarious, and where it was organized from above, it had traces of the sycophantic character evident at Tsinghua.

Chapter Seven

Eliminating the Distinction Between Mental and Manual Labor

In May 1970, a team of workers armed with sledge hammers tore down an elaborate model of channels and flues that had been built by Tsinghua's water conservation department just before the Cultural Revolution to conduct experiments in water flow dynamics. The workers' propaganda team now in charge of the university decided the large hall that housed the model was needed to assemble diesel trucks. These trucks were to be the principal product of a new campus-based factory that was to employ 1,500 people and become the primary focus of several of Tsinghua's engineering departments. The previous summer, most of the faculty of the water conservation department had been dispatched to various sites along the Yellow River. In line with Mao Zedong's education philosophy, Chi Qun, the top leader of the workers' propaganda team, had declared that if the department remained on the university campus in the capital, it would continue to be "isolated from practice" (*tuoli shiji*). Instead, Chi and the university's new leadership decided the appropriate place for the department's work was on the banks of the Yellow River, where it could conduct "open door education" (*kaimen banxue*) and practical research. The department built a branch school near the huge Sanmen Gorge dam in Henan Province, recruiting students from rural communes in several provinces along the river. For the next seven years, classes were held in this branch school as well as at the Tsinghua campus in Beijing, but students and teachers spent much of their time in the field. They took part in efforts to remediate silting problems at the Sanmen Gorge dam and trained commune members to prevent soil erosion, build and maintain irrigation facilities, repair dikes, and construct small dams for irrigation, flood control, and electricity generation. They promoted methods to prevent soil erosion and tried to convince village leaders not to plant crops in mountainous terrain. They instructed villagers how to measure the results of their efforts and use the data to evaluate and improve irrigation, antisilting, and flood control methods.[1]

Although Mao and his radical followers insisted that sending professors from elite universities to help peasants dig irrigation ditches served economic development, they did not hide their primary aim—to prevent the development of a privileged and detached class of experts. Indeed, the radical education policies of the Cultural Revolution were all fundamentally justified as part of an effort to eliminate the distinction between mental and manual labor. In China, class distinctions based on knowledge enjoyed greater legitimacy than those based on either private property or political power. Wealth and political position could be derided as ill-gotten, and claims to social standing and power made based on money or official rank could be challenged, but claims based on knowledge were much less vulnerable. Mencius's maxim, "Those who labor with their minds rule others; those who labor with their strength are ruled by others" (*laoxinzhe zhiren, laolizhe zhiyu ren*), had been reinforced over the centuries by the imperial examination system and was widely taken for granted as natural, inevitable, and just. During the Cultural Revolution, Mao not only directly challenged the truth of the maxim, but sought to render obsolete the categories on which it was based, launching the most thoroughgoing effort to undermine the cultural foundations of class power in Chinese history.

The huge social gap that separated intellectuals from the laboring classes became officially intolerable, requiring immediate, radical solutions. Working people were to master knowledge (*laodong renmin yao zhishihua*), while intellectuals were to become accustomed to doing manual labor (*zhishifenzi yao laodonghua*). The education system was completely overhauled with the aim of eliminating, to the extent possible, its capacity to reproduce class differentiation. The explicit goals were to prevent the reproduction of an educated elite and to undermine its claims to social power. Radical education reforms brought about dramatic changes throughout the school system and gave rise to acute problems. Rapid expansion generated severe problems for schools at lower levels, particularly in rural areas, and the elimination of entrance examinations forced schools at all levels to reorient their teaching goals. Turmoil and change, however, were most wrenching at higher-level schools, which were more dependent on the examination system and which became particular targets of anti-elitist measures.

In China today, the Cultural Revolution decade is officially referred to as the "ten years of turmoil" (*shinian dongluan*). Most Chinese scholarship about education during the Cultural Revolution decade has reflected what might be called the consensus view in the Chinese education establishment—that radical policies disastrously eroded the quality of education. Radicals, it is argued, not only laid excessive stress on expanding basic education, but also displayed a general hostility toward intellectuals and theoretical knowledge. Most studies have focused on the harm done to the upper end of the education system, especially universities and keypoint middle schools.[2] Western scholars who have investigated the impact of Cultural Revolution policies on the upper end of the Chinese education system have come to similar conclusions.[3] Studies about the lower echelons of the education system have been less concordant and many have diverged from this bleak assessment. The emphasis on popularization had a much different impact on those at the bottom than on those at the top, and this has been reflected in quantitative studies that traced sharp gains in educational attainment for rural children and for rural girls in particular.[4] Based largely on extensive interviews with rural teachers, Suzanne Pepper presented a mottled picture of the reality behind these numbers, describing the difficulties encountered in rapidly expanding village schools and the shortcomings in the quality of education provided.[5] She was impressed, nevertheless, that radical policies made middle school education available to the majority of village youth for the first time. Dongping Han argued that the rural-oriented practical training provided by rural middle schools during the Cultural Revolution decade was a key factor that facilitated rural economic development in the 1970s and 1980s.[6]

This chapter, which analyzes the impact of Cultural Revolution education policies at Tsinghua University, joins the literature that has examined the turmoil at the top end of the education hierarchy during this period. The analytical focus, however, is somewhat different from that of most previous scholarship, which has largely been concerned with conventional education goals, that is, the quality and quantity of training provided. Instead, I identify the class-leveling aims that were the fundamental inspiration for Cultural Revolution education policies, analyze the methods employed, and evaluate the results in terms of these aims. After presenting a brief chronological overview of the battles over education policies during the Cultural Revolution decade, I will examine in detail the results of these policies.

Tug-of-War over Education Policy

During the late years of the Cultural Revolution decade, radical and conservative factions in the CCP contended fiercely over education policies, which swung from Left to Right and back again. Extreme radical views prevailed during the first years after schools were closed in 1966. Primary schools and middle schools gradually reopened, but at first, much of the instruction was politically oriented, and all colleges and universities remained closed until 1970, when Tsinghua and a small number of other schools reopened their doors. Most universities did not reopen until 1972, after a marathon four-month national meeting in 1971 finally established new guidelines for university education.[7] The meeting featured a protracted debate between Chi Qun, the leader of the Tsinghua workers' propaganda team, and veteran Tsinghua official He Dongchang about how to evaluate the "Sixty Articles," which had guided higher education policy before the Cultural Revolution. The meeting resulted in a compromise. Chi Qun was able to supervise the drafting of the meeting's summary statement, which roundly condemned higher education policies before the Cultural Revolution and provided a radical general orientation for university education.[8] Nevertheless, the meeting authorized efforts to restore a more systematic university teaching program. Over the next two years, He Dongchang led an aggressive effort to do so at Tsinghua, with Premier Zhou Enlai's support. At first, Chi and the workers' propaganda team at Tsinghua acquiesced to many of He's conventional initiatives, but they soon decided that restoration of past practices had gone too far.

As will be discussed in Chapter 8, a major battle ensued after conservatives won the restoration of supplemental college entrance examinations in the spring of 1973. The radicals ultimately defeated this effort and, encouraged by the ascendancy of radical power at the Tenth Party Congress in the autumn of 1973, the Tsinghua workers' propaganda team launched a major campaign to Oppose Retrogression (*fan huichao*). He Dongchang, who was then in charge of academic programs at Tsinghua, was criticized for restoring elitist practices and demoted to a subordinate position in Tsinghua's mechanical engineering department; many other veteran cadres were also demoted or transferred. In

January 1974, with Mao's support, the radicals extended the Oppose Retrogression campaign into a nationwide movement to Criticize Lin Biao and Confucius. During this movement, their influence reached its apex. At the end of 1974, however, the tide turned once again. Deng Xiaoping was named first deputy premier and authorized to carry out a wide-ranging "rectification" of economic and education policies. Zhou Rongxin, an advocate of more conventional policies, was named to head the newly restored Ministry of Education.

Throughout 1975, the radical and conservative camps energetically pursued contrary education policies. While Zhou Rongxin attempted to use the education ministry to rectify education at all levels, Zhang Chunqiao, who was charged with overseeing education affairs for the State Council, collaborated with Chi Qun and Xie Jingyi to further promote radical education initiatives. In October 1975, the most senior party official at Tsinghua, Liu Bing, wrote a letter to Mao to complain about Chi's authoritarian management style.[9] The letter, delivered by Deng Xiaoping, was presumably intended to erode radical influence in the education field by undermining a key radical leader at his base of operations. Mao, however, rejected the letter and authorized Chi and Xie to launch a campaign against Liu Bing at Tsinghua. This campaign was soon expanded to become a Great Debate on the Education Revolution, and then in early 1976, with Mao's blessing, further expanded into a broader political movement to oppose the Right Deviationist Wind to Reverse the Correct Verdicts of the Cultural Revolution, the main target of which became Deng Xiaoping.

The class-leveling program in the cultural field carried out during the Cultural Revolution decade continued to pursue the three basic goals of the Education Revolution initiated during the Great Leap Forward: (1) increasing political power at the expense of cultural power; (2) redistributing cultural capital; and (3) altering conventional academic credentials and occupational categories in order to combine mental and manual labor. Slogans and practices of the Great Leap Forward years were revived, but this time implementation was more resolute and protracted, and because the examination system was eliminated, the environment was much more amenable to radical practices. The following sections examine the results in various fields. They are organized thematically rather than chronologically, an approach that sacrifices attention to the twists and turns over time in favor of assessing the overall character and consequences of radical policies that were dominant throughout the decade.

Redistributing Access to Education

The agenda of the Cultural Revolution in terms of redistributing access to education was very ambitious. The goal was to level the education pyramid, compressing the top downward, and building up the base, creating a flat system that distributed educational attainment equally across the population. The number of primary and middle schools was expanded at a breakneck pace, especially in rural areas, with the aim of making both primary and secondary education universal in a short period of time.[10] Although regional variation was encouraged, all schools were supposed to be equivalent in terms of quality. The keypoint system of selective college preparatory schools was eliminated, and these schools, along with the special schools that had catered to cadres' and military officers' children or had been reserved for children of university employees, were converted into regular neighborhood schools. As a result, the middle school attached to Tsinghua University was required to open its doors to an influx of graduates from primary schools in neighboring villages. The length of schooling was shortened. In many areas, primary school was cut from six to five years, and middle school education was typically reduced by one or two years. Previously, students had taken entrance exams starting in the fourth grade of primary school to determine whether they were qualified to continue to the next level of education and whether they would attend a keypoint or a regular school. Now, by making primary and secondary education universal and by making all schools nominally equal in terms of quality, there would no longer be a purpose for entrance examinations.[11] All children were supposed to attend school for nine or ten years and then go to work.

In terms of this goal, the results were impressive. Before 1966, less than half of Chinese children completed primary school, only about 15 percent completed junior middle school, and less than 3 percent completed senior middle school. By the end of the Cultural Revolution decade, almost all children completed primary school, over two-thirds completed junior middle school, and well over one-third completed senior middle school.[12] These figures, of course, require qualification: the years of schooling were cut, and education standards had declined.[13] Nevertheless, the rapid expansion of basic education during the Cultural Revolution decade allowed—for the first time—the great majority of Chinese children to complete primary school and attend middle school.[14] The consequences of the Cultural Revolution's social-leveling program in education were, of course, very different at the top. While educational opportunities for rural children were greatly expanded, the opportunities available to children of urban elites were painfully curtailed. Many children from intellectual and revolutionary cadre families had grown up expecting to attend keypoint middle schools and universities, which were now eliminated or fundamentally altered by harsh anti-elitist policies.

Tertiary education presented a difficult dilemma for Mao's cultural-leveling project. Making middle school education universal was a feasible short-term goal; for the foreseeable future, however, only a small fraction of the population would be able to attend college. For this reason, Mao was uncomfortable with the entire idea of university education. Because he was also committed to modernization, he was willing to compromise egalitarian goals in order to allow advanced training for a minority, especially in the fields of science and engineering, but he favored programs that were shorter and had more of a mass character. The compromise solutions he proposed were concisely summed up in his July 21 (1968) Directive. "It is still necessary to have universities; here I refer mainly to colleges of science and engineering," Mao wrote. "However, it is essential to shorten the length of schooling, revolutionize education, put proletarian politics in command, and take the road of the Shanghai Machine Tools Plant in training technicians from among the workers. Students should be selected from among workers and peasants with practical experience, and they should return to production after a few years study."[15]

During the remainder of the Cultural Revolution decade, this directive served as the basic template for tertiary education. The university entrance examinations were replaced by a system in which worker-peasant-soldier students were recommended by factories, villages, and military units. This system, which will be examined in more detail in Chapter 8, took selection out of the hands of the school system and made it a political process that was more concerned with political than with academic qualifications. The recommendation system was designed to radically redistribute access to education, the same ambitious goal that inspired the flattening of the primary and secondary education systems. At the same time, the content of schooling was transformed in order to pursue a goal that was in some ways even more ambitious—to change the nature of knowledge and of labor. The aim was to make all education combine theoretical knowledge and manual skills, preparing students for occupations that would require both. Universities were no longer to produce "the kind of bourgeois intellectual aristocrats who use knowledge as capital and lord it over the people."[16]

Reforming the "Breeding Ground of Intellectual Aristocrats"

Radical education policies during the Cultural Revolution were premised on "two estimates" about education during the seventeen years of Communist rule prior to the Cultural Revolution.[17] First, schools had remained under the control of bourgeois intellectuals, and second, most intellectuals—including those trained after 1949—still retained a bourgeois worldview. Universities, in particular, remained the "hereditary domain of the bourgeoisie" (*zichan jieji shixi lingdi*). Even after 1949, they had retained the "feudal" (traditional Chinese) and "bourgeois" (Western) characteristics of the educational system inherited from the old regime, adding to these "revisionist" educational practices borrowed from the Soviet Union. All three, in Mao's view, had promoted educational elitism. Elite universities were now seen as breeding grounds for a new privileged stratum that was prone to lead the country down the capitalist road. A 1971 article published in *Tsinghua Battle Report* (the new title of the official university newspaper) warned that this was what happened in the Soviet Union. "In the first socialist country, on the education front the bourgeoisie actually exercised dictatorship over proletariat and they continued to cultivate a privileged stratum that would restore capitalism. This is a very serious lesson. The Soviet revisionist chiefs like Brezhnev are typical representatives of the privileged stratum—wearing the titles of 'engineer' and 'expert'—that was cultivated by this kind of education."[18]

REEDUCATING INTELLECTUALS

To the leaders of the workers' propaganda team now in charge of Tsinghua, the Red engineers who had been trained by the university before the Cultural Revolution looked a lot like their counterparts in the Soviet Union. During the first seventeen years of Communist power, it now seemed, all those who had been trained at Tsinghua, including those who had joined the party, had absorbed a bourgeois worldview. This accusation came as a shock to young Communist teachers like Wei Jialing. Wei, the Tsinghua student profiled in Chapter 3, had fully embraced Communist ideals in 1958 while helping build the Miyun Reservoir. Since then, she had joined the teaching staff at Tsinghua and become a party member, and she had continued to strive with all her heart to be both Red and expert. Like many of her young colleagues, she considered herself to be very different from older professors at the university. Although senior faculty members had accommodated themselves to Communist power, many professors had never fully accepted Communist ideas; some had only reluctantly given up the traditional long gowns and Western suits that had distinguished them as intellectuals, and they identified with traditional Chinese and Western culture that the CCP attacked as feudal and bourgeois. Wei, in contrast, completely identified with the new order, but after 1966 that no longer seemed to make any difference. "Before we felt we had to reform the old teachers," Wei recalled. "Now we realized we were the targets of reform."[19] Wei came from an old elite family, but even Tsinghua graduates of more humble origins were no longer trusted. Wang Xingmin, who graduated from the university in 1969 and had been retained to work at the school by the workers' propaganda team, was from a poor peasant family. Nevertheless, he said, "We had all been cultivated by the Old Tsinghua, so we were all the targets of reform."[20]

University education before the Cultural Revolution, radical leaders claimed, had been characterized by the "three separations" (*san tuoli*): separation from practice, separation from the workers and peasants, and separation from productive labor. Teachers and students had been allowed to live a privileged lifestyle and they had come to believe they were entitled to such a lifestyle because of their learning. As a result, the old Tsinghua had served as a "revisionist dye pot," which had changed the color of its graduates. Life at Tsinghua had even compromised students of worker and peasant origin, who had lost their class perspective as they strived to become engineers. This process was described in a three-line verse:

> First year, country bumpkin
> Second year, urbane
> Third year, refuse to even recognize ma and pa[21]

As soon as the workers' propaganda team arrived at Tsinghua in 1968, it enthusiastically set out to eliminate the three separations. In order to reform intellectuals' worldview, challenge their traditional sense of superiority, and free them from any lingering disdain for manual labor and for workers and peasants, intellectuals were sent to the countryside to be "reeducated by the peasants" (*jieshou nongmin zai jiaoyu*). In May 1969, most Tsinghua teachers, cadres, and staff were sent to a "May 7 cadre school" in the southern province of Jiangxi.[22] They arrived at an undeveloped patch of land called Liyuzhou, where they built their own housing and reclaimed marshy lakeside bottomland for agricultural cultivation. Work at Liyuzhou was hard and the conditions primitive, especially for teachers accustomed to living at a top-tier university in Beijing.[23] By the time they returned to Tsinghua in the fall of 1970, when the first cohort of worker-peasant-soldier students arrived, most of the nearly two thousand teachers, cadres, and staff who had been sent to live on the Jiangxi farm had contracted schistosomiasis, a serious parasitic disease that particularly afflicts rural residents in marshy areas.

Many teachers responded with indignation to the abrupt change in lifestyle caused by their transfer to Liyuzhou. Liang Yousheng, the son of a Nationalist official who began teaching at Tsinghua in 1958, recounted how Chi Qun spoke to teachers at the farm as they squatted in the blazing summer sun. "He said we intellectuals couldn't take hardship. Here we were, very educated people wearing nothing but short pants and hats, and that's how he talked."[24] Others remembered the experience with bittersweet appreciation. Lai Jiahua, whose father had been a wealthy factory owner, accepted the thinking behind the requirement that intellectuals live in the countryside. "The old intellectuals looked down on the workers and peasants," Lai told me. "Mao's thinking was that . . . intellectuals should serve the workers and peasants," and they could only learn to do that by "living, eating, and working with workers and peasants." He remembered the months in Jiangxi as very difficult, but not without benefit. "I ran water buffalo to prepare rice fields and I became a very good water buffalo driver. This gave me spiritual strength; it made me realize that I could do anything I dared to do. It also gave me the ability to work very hard. . . . I also got a chance to make friends with peasants. I came from a better-off family and I had no friends who were workers or peasants before."[25]

Instead of going to the Jiangxi farm, teachers in some departments were assigned to help expand factories on the Tsinghua campus or teach villagers about irrigation and small-scale hydropower generation. All teachers and cadres who did not go to Jiangxi, however, were later expected to spend a year doing agricultural labor on a farm that Tsinghua managed in nearby Daxing County. Moreover, even after they returned to the university, cadres and teachers were required to do regular stints of manual labor. Many Tsinghua faculty members were assigned to work in Tsinghua's own factories, where along with conducting classes and doing research they were required to take part in production. "I worked one day a week in production," Hong Chengqian, the Tsinghua-trained manager of one of the university's factories, recalled. "I had specific duties—to run a specific machine at a specific time. When there was an especially big, or heavy, or dirty job, I was expected to take the lead."[26]

UNDERMINING RANKS AND PRIVILEGES

Cultural Revolution policies also diminished material privileges to which faculty members and cadres had become accustomed. After it took power in 1949, the Communist regime distributed goods and services in a much more egalitarian manner than in the past, but some people still received significantly more than others. In the 1950s, all state employees were assigned a professional rank based on seniority and professional accomplishment, and the allocation of wages and other goods and services was largely based on rank. Faculty members were divided into four ranks and workers were divided into eight, with each rank subdivided into grades. In the mid-1960s, workers' monthly salaries ranged from 16 yuan for apprentices to 107 yuan for the most senior and skilled workers, while faculty salaries ranged from 46 yuan for newly hired assistant teachers to 360 yuan for a small number of the most senior professors.

During the Cultural Revolution decade, the ascetic and egalitarian ethics of the insurgent Communist movement were revived. "The idea was that everyone should live at the level of the workers and peasants," recalled Hong Chengqian. In 1971, salaries were increased for the first time since before the Cultural Revolution, but only for those with monthly salaries under 55 yuan, which meant that the great majority of teachers and many senior workers did not get a raise. Most younger workers, on the other hand, received a one-grade wage increase. After 1971, employees received routine promotions and corresponding raises, but the only ones eligible were those at the lower-wage levels. Thus, while Cultural Revolution wage reforms left intact huge disparities, the direction was unmistakable; the range of wages among the younger generation would soon be quite narrow. Among teachers, professional rank became a living ghost. Before the Cultural Revolution, one's place in the rank hierarchy, starting with assistant teacher and progressing through lecturer and several grades of professor, was an elemental characteristic that defined identity, set life goals, established social status, and determined material well-being. Now the hierarchy of ranks was viewed as a violation of Communist principles, and all faculty members were referred to by the generic term "instructor" (*jiaoshi*). Salaries, however, were still based on rank and were now frozen in place.

Privileged access to medical care based on rank also became a casualty of the Cultural Revolution. Like many large work units, Tsinghua provided medical care for its employees and students through a campus clinic, and the clinic sent patients with more complicated problems to Beijing hospitals. In the 1950s, the university had established a special program that allowed the highest-ranking professors and university cadres to avoid waiting in line for care and, if need be, they were likely to be referred to the best hospitals in the city. After 1966, the special program was eliminated, and they waited in line like everyone else.

The most dramatic change, however, was in housing allocation. Before the Cultural Revolution, the university had assigned apartments to employees based on rank. Workers lived in five "workers' districts," composed of blocks of old, cramped, single-story apartments, while teachers and cadres were given new apartments in high-rise buildings, each building constructed to specifications of size and comfort suitable to a specific range of ranks. This method of allocating housing reinforced social segregation by rank. Tong Xiaoling, the daughter of a very successful young professor, described how the high-rise apartment building where she lived in the early 1960s became a closed world. "The residents were all high-ranking teachers or cadres," she told me. "The rent was higher and we had hot running water and sit-down toilets, which at that time were both luxuries. But I didn't know that then, because I didn't know how poorly other people lived. I never went to other classmates' houses; I just went to some of my friends' apartments in the same complex. It was like we lived in a beehive. The kids in these buildings were becoming a new class; we were not like other kids."[27]

After 1966, this kind of segregation by occupation and rank was no longer acceptable. Using a slogan borrowed from the days of Land Reform, "Overthrow the landed tyrants and divide the land" (*dadao tuhao fen tiandi*), the radical student faction presided over the redistribution of university housing. High-ranking cadres and professors were forced to move to smaller, less well appointed apartments, or to share their accommodations, and the vacated apartments were taken over by families of workers and young teachers. Some of the larger townhouses, which had previously housed a single family, were now occupied by three—one upstairs, one downstairs, and one in the former maid's quarters. Tong Xiaoling and her family surrendered their relatively luxurious apartment to a worker's family in 1966. "In the Cultural Revolution, they smashed privilege (*dapo tequan*)," she explained, recalling that her family had to move six times over the next decade. In 1968, the workers' propaganda team ratified the redistribution of housing that had already taken place and then organized an intensive effort to build new apartments. As new housing came on line, workers and teachers were assigned to the same buildings, and priority was determined by family size instead of rank. Since workers typically married earlier than teachers, they were often first in line for new apartments.

Members of the workers' propaganda team were strong proponents of the egalitarian and ascetic policies of the Cultural Revolution and were equally enthusiastic about implementing them. The workers and soldiers who made up the team continued to receive their regular—usually quite humble—salaries from their factories or military units (to which they would return after their stints at the university). Team members—who often had experienced much harsher working and living conditions in their own units—proudly took the lead in projects that involved manual labor and difficult living conditions, especially during extended visits to villages, construction sites, and factories. Their enthusiasm, endurance, and abilities in these areas shored up their authority at the university, helping compensate for their educational disadvantages vis-à-vis university cadres and teachers.

TRAINING "LABORERS WITH SOCIALIST CONSCIOUSNESS AND CULTURE"

Even though the workers' propaganda team had taken control of Tsinghua and its classrooms were now filled with worker-peasant-soldier students, there remained an abiding concern that the university would continue to turn the young people in its care into intellectual aristocrats. The challenge was to impart advanced knowledge to young people without cultivating a class apart. Students were still expected to take the Red and expert road, but they were no longer to become Red experts. This subtle semantic difference reflected a conundrum that lay at the heart of the educational experiments of the Cultural Revolution. Instead of becoming Red experts, students were to strive to remain part of the laboring classes, becoming "laborers with socialist consciousness and culture," a term revived from the Great Leap Forward.

Many of the ideological campaigns conducted during this period were connected with this overarching concern. During the 1974 campaign to criticize Confucius, for instance, two attitudes attributed to Confucian teachings about education were made particular targets. The first was treating "knowledge as private property" (*zhishi siyou*). Members of the educated classes, it was argued, sought to monopolize knowledge and use it for their private benefit; in a capitalist society, college graduates considered what they had learned to be a "commodity" that they could sell on the labor market. They believed that ownership of knowledge gave them the right to live a privileged life. In fact, it was argued, knowledge was the common patrimony of humanity and had originated in the working people's struggle for survival. Students, therefore, should not consider the advanced knowledge they learned in college as private property to be used for their own benefit, but rather as a resource with which to better serve the people. Students were also asked to reject the traditional idea of "studying to become an official" (*dushu zuoguan*). Their goal in going to school, they were told, should not be to rise above the common people, but rather to return to the ranks of the people after studying. The humility that this new generation of college students was expected to adopt was signaled by their title, *gongnongbing xueyuan* (worker-peasant-soldier student). In addition to the proletarian modifier, the term *yuan* means "member," conveying a collective identity parallel to *sheyuan*, the term used to refer to members of rural communes. It had a more modest ring than the term *daxue sheng* (university student), which had been used before the Cultural Revolution.

Life at Tsinghua was designed to inculcate the ascetic Communist ideal of "hard struggle and plain living" (*jianku fendou*). Soon after the first group of worker-peasant-soldier students arrived at Tsinghua in 1970, they were sent out on a 500-kilometer military-style training march in the mountains north of Beijing. They marched 50 kilometers a day, staying in villages at night. The traditions of doing volunteer labor in villages and participating in basic military training were revived with renewed intensity. Open door education policies required that students spend considerable time in factories and other off-campus work sites, living in primitive conditions and engaging in physical labor. Architecture students helped build dormitory buildings and factories on campus and took part in construction projects around the Beijing area, while water conservation students went to study at the Miyun Reservoir, where they first built their own rough-hewn dormitories and classroom buildings and then, between classes, hauled earth to shore up the reservoir's massive dams.

Radical leaders at Tsinghua often charged that before the Cultural Revolution, Jiang Nanxiang and other university officials had dismissed peasant youth as too coarse for university training. The prevailing attitude, they charged, was: "Fine designs cannot be carved on crude pottery" (*cuci chawan diaobucheng xihua*). Although such attitudes no doubt existed, it was unfair to direct this criticism at Jiang, who had enthusiastically welcomed rural students if they met Tsinghua's rigorous academic standards, and, in fact, he had given students from poor and lower-middle peasant families particularly solicitous treatment when it came to political cultivation. Using the same metaphor, the change in Tsinghua's mission after 1966 can perhaps be better described as follows: while Jiang had perfected the process of shaping coarse clay into fine ceramics, Chi and Xie were determined to produce rustic pottery.

Combining Theory and Practice

Cultural Revolution education reformers wanted to leave no refuge for theory that was not combined with practice. The radicals articulated a variety of philosophical and pedagogical arguments about the advantages of practical learning, but these were always linked to class-leveling goals. By detaching theory from practice and making a fetish of abstract theory, they argued, intellectuals had created fields of knowledge they could monopolize, enhancing their own status and power. Cultural Revolution policies and discourse were designed to adjust the relative value assigned to theoretical and practical knowledge, undermining the currency of "book learning." The knowledge of a professor who could explain the principles of internal combustion but did not know how to fix a diesel engine was now considered to be seriously deficient, and such one-sided learning became the subject of ridicule. Practical efforts to "combine teaching with research and production" were intensified. On the one hand, university campuses were physically transformed by making them into centers of production as well as research and teaching; on the other hand, school leaders endeavored to "free higher education from the confines of the university."

"FACTORYIZING" TSINGHUA

As an engineering school, Tsinghua was in a good position to combine education with research and production, and the workers' propaganda team made a priority of realizing the Great Leap Forward goal of "factoryizing" (*gongchanghua*) the university. After the propaganda team took control of Tsinghua in

1968, it greatly expanded the factories and practical research centers that had been established before the Cultural Revolution, including the machine tools factory, the electronics factory, the equipment factory, the precision instruments factory, the computer factory, and the experimental nuclear reactor. It also built several new factories, although its grandest initiative—the truck production plant—was scuttled as misguided in 1972. The team made sure that Tsinghua's factories carried out production, but after the demise of the truck factory, they conceded that the main function of campus-based factories was to support research and teaching. The expansion of Tsinghua's factories dramatically increased the proportion of university employees who were workers (as opposed to intellectuals). Hundreds of demobilized soldiers and middle school graduates were hired, and the number of employees in campus production facilities grew from 626 in 1966 to 1,841 in 1977.[28]

As part of the effort to combine teaching, research, and production, institutional links were established between universities, research institutes, and factories. Tsinghua benefited from this reorganization as funding, equipment, and researchers were transferred from research institutes to Tsinghua's factories. Some of these factories were given responsibility for well-endowed national priority research and development projects involving computers, electronics, and nuclear technology. These projects were supported by both radical and conservative leaders at the center, including Jiang Qing and Zhou Enlai, and by both the workers' propaganda team and leading scientific cadres at Tsinghua. On the one hand, the projects were consistent with the radicals' insistence on combining teaching, research, and production, and they allowed the propaganda team leaders at the university and their radical patrons at the center to enlarge the size and prestige of their Tsinghua domain. On the other hand, they were consistent with conservative leaders' efforts to promote technological development.

Tsinghua was restructured under the slogan, "factories lead academic disciplines" (*gongchang dai zhuanye*), and where practical, campus factories and departments were merged (*changxi heying*). For instance, Tsinghua's machinery factory was combined with the mechanical engineering department to form a machinery factory/department. A leadership committee was drawn from both sides and several rank-and-file factory workers were selected to participate on this committee. Majors within the department were administratively linked with related shops in the factory, teachers were assigned to work in the factory, and factory technicians and workers were assigned to teach classes.

Although radical education practices, including this kind of obligatory interaction between intellectuals and workers, were disconcerting to faculty members, many accepted the virtue of the Education Revolution and they did their best to understand and carry out its requirements, even when these requirements appeared eccentric or impractical. Even those who had deep reservations accommodated themselves to radical policies. Most, like Hong Chengqian, the Tsinghua graduate who served as a manager in one of the university's factories, were ambivalent. On the one hand, Hong believed that the close links between academic departments and factories had a positive impact on design and production. "During the Cultural Revolution, creativity was very strong," he told me. "The main reason was that teachers brought a lot of knowledge to the factory." Collaboration between teachers and workers, Hong noted, led to advances in designing new products. "[The workers] helped in design," he recalled. "They made suggestions about how to improve the products, how to simplify the production process, making it more efficient and more practical." On the other hand, he thought many Cultural Revolution policies were ill conceived. "Mao . . . wanted a proletarian university, a Communist university, [he wanted] to combine mental and manual labor and combine book learning and practice—for people to be both teachers and laborers." These might have been commendable goals, Hong said, but they were impractical. "The nature of the work of the two groups—workers and teachers—is not the same; they have different qualifications and their quality (*suzhi*) is not the same," he explained. "The workers' cultural level is relatively low and the teachers' cultural level is higher. The idea during the Cultural Revolution was to put them both together, to make them be together, to make teachers participate in production and workers participate in teaching. But what the workers talked about, the teachers weren't interested in, and what the teachers talked about, the workers weren't interested in. . . . Teachers and workers didn't have a common language. . . . The thinking of the Cultural Revolution was to get rid of the differences between the two groups, but that was utopianism."[29]

OPEN DOOR EDUCATION

The ideas of open door education and of combining teaching with research and production, of course, were not foreign to Tsinghua faculty. The Soviet education model adopted in the 1950s had stressed practical training, and during the Great Leap Forward students and teachers had been exhorted to closely link theory and practice. Before the Cultural Revolution, however, practical training had been largely confined to the last stage of a three-stage pedagogical approach, which started with basic courses (mathematics, basic sciences, and politics), then advanced to courses related to specific majors, and finally provided practical training. During the Cultural Revolution, radical leaders denounced this "old three-stage" policy as "giving priority to book learning" (*zhiyu diyi*). Starting with abstract theoretical courses, they argued, alienated worker-peasant-soldier students, who were more adept at grasping theory if it was combined with practice. The new curricula were designed to combine all three stages and eliminate the "teacher-centered, classroom-centered, and textbook-centered" approach of the first two stages.

While the workers' propaganda team insisted on limiting the hours spent teaching abstract theory, teachers felt that certain basic math and science courses were essential. Because most students had not completed senior middle school, curricula had to start with middle school math and science and then teach university-level content, all in a program that had been cut from six to three and one-half years. A tremendous amount of material was crammed into accelerated courses, and the result, especially during the 1972–73 period when He Dongchang was in charge of academic programs, was highly pressurized book learning. Despite exaggerated expectations reminiscent of the Great Leap Forward and the earnest efforts of teachers and students, many of the students, especially those who had not completed senior middle school, were simply unable to master so much material.

Following the radical turn in 1973, the open door approach once again came to the fore. In many majors, teaching was linked to research and production in Tsinghua's own factories. In one half-work/half-study experiment, students and teachers were assigned to a computer production facility in the university's electrical instrument workshop, where they joined with workers to form groups that worked and studied together. Not only were the students and teachers working in production, an article in the university newspaper reported, but also within two years, 40 percent of the workers were expected to take on teaching responsibilities. "Through 'intellectuals becoming accustomed to manual labor' and 'workers mastering knowledge,'" the article concluded, "we are creating a better situation in which workers will grasp scientific technique and genuinely participate in the Education Revolution, opening up a new road and greatly encouraging worker comrades' enthusiasm."[30] Majors that did not have appropriate factories on campus developed relationships with other work units. Students and teachers relocated to factories for months at a time, "living, eating, and working together" with the workers. In one common pattern, they did manual labor and studied production processes during the day and attended classes at night. These classes were mainly taught by faculty members, but factory technicians and workers were also invited to give lectures, as were students. In their second and third years of study, students and teachers were asked to tackle production problems. For their graduation projects, small groups composed of several students, a teacher, and a factory technician would be assigned to solve a particularly complex problem.

Despite its egalitarian rhetoric, Tsinghua remained a very elite institution even in its worker-peasant-soldier student incarnation. Because of the special attention it received from Mao and central party authorities, as well as the special research responsibilities it was given, Tsinghua's superior standing, in comparison with other schools, increased. The university retained its highly qualified teaching staff and superior facilities, and—as a national-level university—it enjoyed first pick among the young people recommended for college admission. The university's academic courses were simplified to such an extent that later critics would claim—with justification—that it was reduced to a glorified vocational school. But it offered elaborate practical training programs very different from those typically provided by vocational schools, which generally occupy the lowest rung of the school hierarchy in conventional education systems. During the Cultural Revolution decade, Tsinghua became an elite experimental vocational school that in many ways provided exceptional training.[31] This onsite training involved great costs, which were only partly offset by the labor and technical services students and teachers provided to the factories that became their classrooms. Both teachers and students stressed that this kind of practical training was much more elaborate than what had been offered before the Cultural Revolution, and it largely disappeared in the reform era. "Now factories won't let students go and work," explained veteran teacher Zhuang Dingqian. "What do they say? 'I have to make money; I have production to carry out.'"[32]

Tsinghua students and teachers who participated in open door education expressed mixed opinions about it. On the one hand, all the students I interviewed praised the hands-on training they received. Luo Jinchu, for instance, was very proud that he and his classmates had worked with senior researchers to design the first generation of minicomputers mass-produced in China. "I think we were very lucky," he told me, reiterating one of the pedagogical principles of the Cultural Revolution decade. "We didn't start from the basics; we started from the object—from the goal—to discuss the basic principles. If you just discuss the principles without understanding the practice, you can't really understand it." Luo recounted how he had invented a method of testing memory chips and he recalled the dedication of his team members. "Even when we dreamed, we dreamed about computers and software."[33] On the other hand, Luo lamented their limited theoretical training. Lai Jiahua, a leader of the computer research program at Tsinghua, was also proud of the computers they designed in the 1970s, which he stressed were completely Chinese creations with no technical assistance from abroad, and he noted that students had great opportunities to learn practical design skills. Nevertheless, because the workers' propaganda team opposed theoretical training, he said, the students did not learn enough to become fully competent engineers and scientists unless they had an opportunity to take graduate classes later. "We had a terrible loss in the pipeline of knowledgeable people."[34]

Eliminating Examinations and Reinforcing Collectivism

Chinese schools not only eliminated entrance examinations, but also attempted to reduce all written tests to a minimum. Radical reformers objected to written tests for several reasons. First, children from educated families were more proficient in written tests; second, written tests encouraged scholastic rather than practical learning; third, examination-oriented education promoted rote memorization rather than creative inquiry and critical thinking; and finally, individual exams stimulated individualism and undermined the collectivist ethic. The alternative selection and assessment methods employed during the Cultural Revolution decade promoted collectivism in ways that the examination system could not. Before the Cultural Revolution, students who tested into Tsinghua were justifiably proud that they had made it to the pinnacle of the education system by their own efforts, and their performance on course examinations at the university continued to spur a sense of individual accomplishment. The recommendation system, in contrast, used a collective mechanism of selection, which was geared to select those who had a particularly strong sense of responsibility to the collective. Students felt a debt of gratitude to the work unit that recommended them, and their diligence in study was ensured not by individual exams, but by the collective monitoring and support of their classmates.[35]

Luo Yaozong, the son of a coal miner who was recommended to attend Tsinghua in 1972, told me that his classmates studied hard because they felt an obligation to the people who sent them to study. "That period was very particular. . . . Everyone had been sent by a different work unit. They had to study well; after they went back, they had to account for themselves (*jiaodai*) and repay their debt (*huibao*). That gave you a lot of pressure."[36] Among the criteria used to select worker-peasant-soldier students, none was stressed more than a commitment to "serve the people," and, according to Zuo Chunshan, a student of peasant origin, he and his classmates felt obliged to show that they were willing to live up to this expectation. "Young people today can't understand it, but that's the way we studied. Our sense of responsibility was very strong. It wasn't just ordinary; it was very strong."[37]

Tsinghua revived experiments with alternative forms of student evaluation that had been introduced during the Great Leap Forward. Open-book exams were encouraged; questions were distributed in advance and students were allowed to work in groups and discuss the answers. Written exams were supplemented with experiments and oral reports.[38] The practical orientation of the curricula was amenable to hands-on methods of evaluation, and coursework was capped by graduation projects in which groups of students collectively tackled practical problems in factories and other work sites.

Collective monitoring by classmates reinforced students' conscientiousness. "Then, there was a group pressure; now maybe there is an individual pressure," explained Zuo. "Today if you don't study, nobody cares; now there are exams, they test you—you can fail the exam. Then, even though there were no exams, if you didn't study hard, everybody would criticize you."[39] Zuo, who was elected to head one of the small groups in his class, recounted how these groups exerted pressure on members to study hard, inquired about problems, and provided support for those who were falling behind. Collectivist principles also governed the radicals' approach to problems created by students' very uneven academic preparation. At first, workers' propaganda team leaders acceded to teachers' requests and allowed them to divide students into faster and slower groups. After the radical turn of 1973, however, the propaganda team insisted on keeping all students together and encouraged teachers and faster students to help those who were having trouble. Zhang Cuiying, another student of peasant origin, explained that each small group in her class included older workers, who had more practical experience, and young villagers like herself, who had less practical experience but were quicker in academic study. "We studied together and we worked hard not to let anyone fall behind. We said: 'Don't let a single class brother fall behind the group.'"[40]

The collectivist methods of the Cultural Revolution decade also encouraged individuals to conform to the group. The recommendation system encouraged young people to conform to the expectations of their work unit collectives, and the collective monitoring and support systems at the university encouraged students to conform to the expectations of their classmates. This type of conformism can be contrasted to the type of conformism that had been encouraged by individual written examinations. These examinations compelled students to recognize correct examination answers as the truth. These answers oriented classroom teaching and promoted rote methods of learning and thinking. Thus, the incentives of the collectivist approach and of individual examinations both tended to dampen independent thinking, but in very different ways.

The July 21 Road

The spirit of Mao's July 21 Directive was reflected in the *nalai naqu* (return to whence one came) principle, according to which worker-peasant-soldier students, after completing their studies, were supposed to return to the work units that had recommended them. In practice, however, it was difficult to make *nalai naqu* the main principle guiding the assignment of Tsinghua graduates. While many factory workers did return to their original factories, very few villagers returned to their rural communes. Most had been trained in industrial technology and their skills were needed elsewhere. The radicals were clearly disturbed by this reality and in 1975 and 1976 they carried out a new campaign to promote the *nalai naqu* principle. In May 1975, *Red Flag*, the most influential party journal, published a report that elaborated the radicals' concerns, insisting that enrollment and job assignment policies must be changed in order to "destroy the 'ladder' leading to 'officialdom' and clear the soil that breeds revisionism in the field of education."[41] The Tsinghua newspaper and the national press subsequently publicized requests by a number of Tsinghua graduates to return to their own rural communes.[42] The press reports heralded this as a new "trend," but this characterization was contradicted by the numbers presented. In 1975, only twelve out of 1,800 Tsinghua graduates "realized their wish to go to the village and work as peasants."[43]

At the same time, radical leaders promoted the development of shorter-term technical training programs that were directly connected to factories and rural communes and were in a better position to put the *nalai naqu* principle into practice. They developed two models. Factories were encouraged to establish "July 21 universities" modeled after the training program created by the Shanghai Machine Tools Plant, which Mao had lauded in his July 21 Directive. These schools provided full-time or part-time classes for workers, preparing them to do technical work in their own factories. In the past, this work had typically been performed by graduates of technical middle schools, colleges, and universities. Because selection had been carried out by the education system (and was largely based on academic criteria), schools had served as a direct ladder to advantageous class positions, facilitating the reproduction of the educated elite. Now a factory's own workers would be trained to do technical work. This was intended not only to impede elite reproduction, but also to break down the division of labor between manual and mental occupations, and facilitate the involvement of workers in managerial and technical responsibilities.

Rural counties were encouraged to establish "May 7 agricultural universities" modeled after the Chaoyang Agricultural University in Liaoning Province.[44] These schools provided short-term courses for local peasants, who returned to their communes after graduation. Previously, graduates of urban-based agricultural colleges and universities had typically been assigned jobs at the county level and above, all of which were located in cities and towns. Now, the Chaoyang model encouraged "going up and down several times" (*jishang jixia*), that is, peasants attended short-term training programs, returned to their communes, and later went back to school for further training.[45]

In the spring of 1975, Tsinghua University began to develop—alongside its three-and-one-half-year regular university program—new short-term industrial and agricultural schools for local workers and peasants. A Spare-time University located on the Tsinghua campus admitted about 1,500 students from 140 factories in the Beijing area. These factories, which did not have sufficient resources to develop their own July 21 schools, included Tsinghua's own factories, as well as small- and medium-sized factories in urban districts, and new rural factories established by communes on the outskirts of Beijing. There were no age restrictions, and about 30 percent of the workers recommended by their factories had already been on the job at least ten years. They studied ten majors, including machine tool principles, automobiles, welding, electrical work, hydraulic machinery, thermal treatment, casting and forging, metallurgy, automation, and environmental protection. The programs varied from six months to two years, and workers typically attended class two evenings and one afternoon a week. Some students were expected to help establish July 21 programs in their own factories after they graduated. In addition, Tsinghua students and teachers were sent to conduct technical classes in factories in Beijing and Shijiazhuang.[46]

That same spring, Tsinghua established a Village Branch School on the university's farm in nearby Daxing County. The first five hundred students were recommended by communes in Daxing and several nearby counties. The school established ongoing relationships with a large number of communes, and as one group of graduates returned, the communes sent a new group of students. Some of the students had already been working as bricklayers, electricians, tractor drivers, and rural technicians, and some were sent-down urban youth (who, by accepting recommendation, were committing themselves to "establish roots" in their new commune homes). Four majors were offered—agricultural machinery, rural electricity, rural water conservation (irrigation and water control), and rural construction.[47] Lu Baolan, who taught at the Village Branch School, reported that her students generally had a low level of education, but were very serious about learning. "It was all relatively simple, but they studied very conscientiously," she told me. "They were all very sincere, very good."[48]

In the summer of 1975, the Tsinghua newspaper proudly reported that new students in the university's short-term popular training classes already outnumbered the new worker-peasant-soldier students entering the university's regular program. This benchmark, the article declared, indicated that the university was making progress in accomplishing its new mission, which was to "serve the workers, peasants, and soldiers; go to the workers, peasants, and soldiers to popularize education; and start with the workers, peasants, and soldiers in raising the level of education" (*wei gongnongbing fuwu, xiang gongnongbing puji, cong gongnongbing tigao*).[49] While before the Cultural Revolution, Tsinghua's leaders had celebrated the university's selectivity, its new leaders declared that the university's goal was now to "go further and further down and get bigger and bigger" (*yueban yue xiangxia, yue ban yue da*).

By 1976, there were 564,000 students in regular university programs nationwide, and the number of regular students admitted that year surpassed the number admitted in any year since the Great Leap Forward. By that time, however, this number was dwarfed by the 2,629,000 students enrolled in thousands of factory-based July 21 universities and rural May 7 agricultural universities. Other types of adult education were also expanding rapidly. By 1976, there were 30,521,000 adults enrolled in basic literacy classes; 127,302,000 enrolled in adult primary schools; and 3,252,000 enrolled in adult secondary schools. All of these programs were far larger than at any time before or since.[50]

As critics of Cultural Revolution policies have pointed out, the so-called July 21 and May 7 universities were hardly universities in the conventional sense, since most of their students had not completed senior middle school. Nevertheless, the radicals insisted on calling them universities, and articles in Tsinghua's newspaper deliberately called the students enrolled in its Village Branch School "from-the-commune-to-the-commune *university students*" (*she-lai shequ daxuesheng*), even while it preferred the more modest term, *xueyuan*, to refer to the university's regular students. The clear purpose was to dilute the meaning of the term *daxue* (university), an intention made explicit by the play on words in the slogan, "*daxue jiu shi dajia lai xue*" (university means everyone comes to study).[51] Semantics aside, there remained real distinctions between the longer-term, higher-quality training received by regular students at Tsinghua and the training received by students at the July 21 and May 7 schools, and the differences continued to be reflected in distinct job allocation policies. The intentional creation of semantic ambiguity signaled the radicals' dissatisfaction with the status quo, and their intention to continue to erode the elite status of tertiary education. The direction of their policies was consistent with a radical vision in which the entire education system would be virtually flat and would be responsible only for dispensing knowledge, not for social selection. All children would complete middle school and then go to work, and further education would increasingly take the form of short-term in-service training.

Leveling the Cultural Foundations of Class Power

In 1976, Han Lingzhi, a young graduate of Peking University's attached middle school who was living in a nearby rural commune, was selected to teach at Tsinghua's attached middle school. Although she was eager to return to the city after a three-year stint in the countryside, she was disappointed with the job assignment. "I wanted to work in the factory as a worker, I didn't want to teach," she explained. "Nobody wanted to teach. To be a worker was a higher class. At that time, a teacher was not considered a good job; a teacher was an intellectual." Han's father was a doctor and she told me she had always been discriminated against in school because she was from an intellectual family. "People from a worker's family or a peasant's family or the army or something, they were superior to us," she recalled. "At that time, it was easier if you had a better occupation—at that time 'better' means worker. . . . It even just sounded better."[52] Han's final sentence speaks volumes about fledgling changes in the status order that the Cultural Revolution was bringing about in China. The fact that it sounded better to be a worker than an intellectual reflected a remarkable change in a society in which freedom from manual labor had been the quintessential mark of refinement, success, and prestige. In a testament to the power of Cultural Revolution class-leveling efforts, this child of privilege was intent on abandoning her intellectual origins to become a worker. The change, of course, was only fleeting, as her remarks would seem as incongruous in China today as they would have before the CCP came to power in 1949.

The goals of the Cultural Revolution in the cultural field were extremely radical and extraordinarily ambitious. Mao and his radical followers, who were concerned that the continued reproduction of the educated classes and the rise of a technocratic elite threatened to extinguish the Communist class-leveling project, converted the long-standing Communist slogan about eliminating the differences between mental and manual labor into a program of immediate action. In their hands, combining theory and practice meant leaving no refuge for purely intellectual occupations, and they were suspicious of any kind of elite training. Not only were their goals radical, but the means they employed were often deliberately excessive. Radical leaders were guided by the logic of Mao's motto, "Proper limits have to be exceeded in order to right a wrong, or else the wrong cannot be righted."[53] Following this logic, excesses were not simply an unavoidable by-product of struggle, but necessary for success. To break down class hierarchies in the cultural field, nuclear physicists had to sweep floors, learn how to slaughter pigs, and be reeducated by peasants.

The fundamental aim of Cultural Revolution education policies was to level class differences in the cultural field, and—in terms of this goal—the practical results were impressive. The education pyramid was transformed into a much flatter structure, and as a result, educational attainment was distributed far more equally. The length of primary and secondary education was compressed and participation was greatly expanded. Tertiary education was changing in the same direction—shorter programs, less sophisticated content, and wider participation. Efforts to make all schools equivalent in terms of quality produced dramatic results, especially at the top. The expansion of basic education mainly benefited children at the lower echelons of society, while the compression of the top of the school system disproportionately hurt children from elite families. These structural changes, together with class line preferences and the replacement of examinations with recommendation, significantly reduced disparity in educational attainment and severely disrupted the reproduction of the educated elite.

Curricula were redesigned to combine theoretical knowledge and manual skills, preparing students for occupations that would require both. Efforts to adjust the relative value assigned to theoretical knowledge and practical skills (increasing the prestige of the latter at the expense of the former) had a real—if temporary—impact on popular conceptions and on the relative social status of intellectuals, workers, and peasants. Symbols of cultural refinement and freedom from manual labor were transformed from marks of social status into badges of disrepute. Intellectuals found themselves the targets of highly intrusive campaigns to lower their social status and compel them to integrate with the laboring classes. The requirements of open door education dramatically changed teaching and daily life at schools. At Tsinghua, campus factories were placed at the center of the educational program, and teachers and students also spent much of their time living, working, and studying at off-campus work sites. Intellectuals participated in manual labor, and not just in symbolic exercises, but for extended periods and on a regular basis. The physical separation of the educated and uneducated classes was reduced, as were differences in their material living conditions.

The extent of change, of course, should not be exaggerated. Cultural capital remained very unevenly distributed and schools still varied greatly in quality. Despite the handicaps imposed on them, children from educated families still enjoyed important advantages in the education system. The occupational structure continued to feature a sharp division between mental and manual labor, and preferences granted to mental labor endured. Nevertheless, an examination of China's tumultuous reality in the mid-1970s reveals important changes in each of these areas. The Cultural Revolution had significantly diminished the class hierarchy in the cultural field, at least temporarily.

Chapter Eight

Worker-Peasant-Soldier Students

On August 29, 1970, over 1,300 worker-peasant-soldier students carrying flags and banners marched through the west gate of Tsinghua University, where they were welcomed with gongs and drums. They were part of the first regular cohort of students admitted to the university since 1965 and they were very different from their predecessors.[1] The students who arrived in 1965 had, with few exceptions, all received top scores on the national university entrance examinations, almost all had graduated from highly competitive keypoint senior middle schools, and most came from China's very narrow strata of educated families. Students in this new cohort had not sat for examinations, but had instead been recommended by their factories, villages, and military units. Most of them had only been to junior middle school and some only had a primary school education, but many had distinguished themselves as model workers or as activists during the Cultural Revolution, and more than half were already members of the Communist Party. While many of the students who attended Tsinghua before the Cultural Revolution had favored leather shoes and Western-style button-down shirts, most of the students who arrived in 1970 wore the cotton-soled shoes and work clothes ubiquitous in Chinese villages and workshops, and as they marched onto the Tsinghua campus one contingent carried pickaxes on their shoulders, symbolizing their class credentials.

The replacement of college entrance examinations with a system of "mass recommendation" (*qunzhong tuijian*) was the most important—and the most controversial—of the education reforms of the Cultural Revolution decade. Radical education reformers criticized the examinations, which had been the lynchpin holding together the entire education system before the Cultural Revolution, for reproducing the educated elite by favoring selection of their children, and for promoting scholastic and stultified teaching oriented to the exams. Eliminating the exams was seen as the key to transforming an education system that had been organized around selecting a minority of talented students for advanced training into one that provided a modicum of practical education for everyone. The fundamental goal of the recommendation system was to remove as completely as possible the responsibility for social selection from the education system. Schools were only to impart knowledge; they were not to select those who would receive further training and promotion. Instead, selection was made a political process, accomplished through political deliberation in factories, communes, and military units.

During the late years of the Cultural Revolution decade, all middle school graduates were assigned to work in a rural commune, an urban work unit, or a military unit, and they became eligible to be recommended for postsecondary training only after working at least two years in a manual labor occupation. Candidates for recommendation generally were required to have completed junior middle school, to be healthy, and to be no more than twenty-five years of age. Although both political and cultural criteria were considered, the emphasis was clearly on the former. The political qualifications were essentially the same as those considered in Youth League and party recruitment, and involved family origin and personal performance. Discussion about the relative qualifications of candidates who met the other criteria primarily involved evaluating individual performance in terms of Communist political and moral ideals. The prescribed selection process involved three stages: "Masses recommend, leaders approve, schools review" (*qunzhong tuijian, lingdao pizhun, xuexiao fushen*). Quotas were passed down the administrative hierarchy to basic work units, mainly communes and factories. The process was supposed to work roughly as follows. First, individuals applied and members of each factory work group or rural production team discussed applicants' qualifications and made recommendations. These recommendations were passed up through the factory workshop or rural production brigade, and on to the factory or commune leadership, which approved the candidates and narrowed the list. The dossiers of the approved candidates were then passed on to the county level. At this level, candidates were assessed to assure that they met minimal educational requirements, and teachers from various colleges and universities converged to select candidates to fill their schools' quotas. The process varied greatly from place to place and changed over time, and every aspect was subject to negotiation, contention, and corruption.

The system was plagued by intractable difficulties and it became a key point of contention between radical and conservative factions in the CCP; in fact, no other issue so sharply defined the epochal factional struggle of the final years of the Mao era. Since its demise, most academic and political discussion of the system has addressed its negative impact on the academic qualifications of college recruits, the problem cited by education officials as the reason for abandoning recommendation in favor of examinations in 1977.[2] While the question of academic qualifications will be treated in this chapter, my main purpose will be to assess the results of the recommendation system in terms of the key goal of its creators—to prevent the education system from reproducing class distinctions. I will consider class distinctions based on both cultural capital, which was favored by the examination system, and political capital, which came to the fore under the recommendation system.

Political Battles over the Recommendation System

Recommendation policies and practices were established only gradually. In 1970, when the first worker-peasant-soldier students were recommended to attend Tsinghua and a small number of other schools, there were few policies in place to involve the masses in the recommendation process. In practice, officials at the factory or commune level, or above, simply chose individuals who could easily be identified as outstanding activists, such as model workers or peasants who had been selected to represent their villages in county conferences to study Mao Zedong Thought. University recruiters also played a key role in the selection of this first group of students. Rank-and-file peasants and workers often knew nothing about the process until after it was over. This was the case with all four members of the 1970 cohort of worker-peasant-soldier students at Tsinghua University who I interviewed.

The next major round of university student recruitment took place in 1972. This time, the number of students recruited was much larger and the widely publicized recruitment campaign was accompanied by new guidelines and a series of articles in the national press exhorting local cadres to involve the masses in the recommendation process. A March 1972 article in *People's Daily* described the model experience of one locality in learning how to carry out recommendation. "At the beginning some comrades thought that ideas could be easily centralized and work would progress quickly if students for admission were nominated by the leadership. They were afraid that opinion would be divided and be difficult to unify if the masses were told to discuss the nominations." Where the leadership simply chose a few candidates without involving the masses, however, the outcome was not good. "Neither the chosen candidates, nor the masses were satisfied. Later they changed their method. They seriously roused the masses, explained to them clearly and thoroughly the meaning, conditions, and methods of student recruitment and urged them to hold discussions and persuade the comrades who met the conditions to apply for admission."[3]

Although practices varied widely, it is clear from the experience of individuals I interviewed that starting in 1972 workers and peasants in many factories and villages were involved in the recommendation process. Nevertheless, by all accounts, local cadres continued to play the central role and many abused the system to secure opportunities for family, friends, and favorites—a practice known as "taking the backdoor" (*zou houmen*). The problem of backdoor admissions plagued the system from its inception and was a matter of great consternation for both opponents and proponents of recommendation. The first rounds of recommendation also produced cohorts of students with widely disparate levels of academic preparation. In 1970 and 1972, efforts to gauge candidates' academic qualifications involved little formal assessment. Recruitment teams sent by Tsinghua, for instance, simply interviewed recommended candidates and asked them to solve several math problems in order to try to evaluate their educational background and potential.

The problems engendered by the recommendation system led conservatives and radicals to propose sharply divergent solutions. Both factions were concerned about curbing backdoor admissions; in addition, the conservatives were intent on raising the academic qualifications of college recruits. The conservatives proposed reintroducing national academic examinations to address both problems. The radicals, in contrast, were committed to mass recommendation and they opposed reintroducing examinations. Instead, they attempted to rein in backdoor admissions by launching a movement against cadres' abuse of power. Radical leaders, who had risen to prominence by attacking the authority of party officials during the early years of the Cultural Revolution, saw the recommendation system as one means to curb local cadres' power by institutionalizing mass political participation.[4]

In her insightful analysis of contention over university admissions policies before the Cultural Revolution, Susan Shirk highlighted three competing principles for selecting college students: *meritocratic*, *virtuocratic*, and *feudocratic*.[5] Meritocratic selection, which employed academic examinations, followed rational bureaucratic principles; virtuocratic selection, which employed political/moral criteria, was typical of charismatic movements; and feudocratic selection, which employed family origin criteria, was more in line with traditional patriarchal principles. Although Shirk's study was limited to the pre–Cultural Revolution years, she suggested that during the Cultural Revolution decade Mao and his radical followers sought to increase the role of virtuocratic criteria, reduce the role of meritocratic criteria, and limit the role of feudocratic criteria. Her typology is useful in analyzing contention over the recommendation system during the Cultural Revolution decade, and the battles between radical and conservative factions in the CCP over recommendation offer evidence to support her suggestion. The radicals stressed political/moral criteria in recommending students, while the conservatives attempted to reintroduce rational bureaucratic methods of selection by incorporating examinations in the recommendation process. Both sought, in different ways, to restrict the corrupting influence of family connections—and social connections, in general—in the selection process. Of course, in both factions there were individual cadres who used their power to favor family and friends, but in principle, each faction saw this practice as a corruption of the principle of selection it favored.

In April 1973, conservatives were successful in reintroducing a set of national qualifying examinations. The examinations, which candidates took after they were recommended by their work units, were similar to the pre–Cultural Revolution college entrance examinations, although standards had been lowered to include questions at both the junior and senior middle school level. The reintroduction of examinations had a number of results that disturbed the radicals. Announcement of the examinations spurred intensive preparation, both in middle schools and among young graduates, reminiscent of pre–Cultural Revolution examination cramming. Work units were induced to stress academic rather than political qualifications in order to recommend candidates who had the best chance of succeeding in the examinations. Finally, the examinations led to hostile standoffs between university representatives and local officials. Several interviewees recounted that most of the top-scoring candidates in the examinations in their rural counties were sent-down urban youth. University representatives insisted that examination scores serve as the final arbiter, while local officials demanded that local youth be admitted as well, despite lower scores.

The radicals denounced the new examinations, and promoted as a hero a youth in Liaoning Province, Zhang Tiesheng, whom they praised for turning in a blank exam inscribed with a note claiming the examinations gave unfair advantage to bookworms who spent their time cramming rather than dedicating themselves to collective work.[6] The radicals' arguments were concisely stated in a September 1973, *People's Daily* article. "It is necessary to carry out appropriate tests and it is wrong to totally ignore educational qualifications, but admissions criteria must conform to the policy of putting proletarian politics in command and giving priority to political qualifications," the author wrote. "We must fully consider practical experience, especially the long-term performance of the candidate in the three great revolutionary movements [production, scientific experimentation, and class struggle], rather than simply admitting candidates according to test scores." The article went on to criticize tests that put "book knowledge above all else." Tests, the author insisted, should not gauge how well candidates had memorized middle school textbooks, but instead should evaluate their ability to use basic knowledge to solve practical problems.[7] The radicals claimed that proponents of the national examinations ultimately wanted to bring back direct enrollment into college of middle school graduates and reestablish the keypoint system of college-preparatory primary and middle schools, which—evoking its graduated structure—had been called the "little pagoda" (*xiao baota*). "By carrying out the bourgeois educational practice of 'selecting the top quality,' they will lure students to climb the 'little pagoda' staircase that leads to becoming bourgeois intellectual aristocrats."[8]

The radicals won this round, forcing cancellation of the new national examinations. To rub in their victory, in late 1973, Chi Qun and Xie Jingyi, the radical leaders of Tsinghua's workers' propaganda team, organized a surprise examination for college professors in Beijing. They invited 613 professors from seventeen Beijing area colleges to participate in a "consultation meeting," announcing only after they had arrived that they were to take a series of examinations similar to those that had been given to worker-peasant-soldier student candidates the previous summer. The results delighted the radical leaders and were widely publicized: two hundred of the professors turned in blank exams, and of those who filled out the answer sheets, over 90 percent received failing scores.[9]

After defeating the effort to reintroduce national examinations, the radical faction moved to shore up the recommendation system by organizing a political campaign against cadre abuse of the system. On January 18, 1974, they launched a campaign against taking the backdoor by publishing on the front page of the *People's Daily* a letter written by the son of a high-ranking military officer who withdrew from Nanjing University after admitting that his father had pulled strings to win his admission. Zhong Zhimin started by confessing that he initially thought it was all right to take advantage of family connections. "I always used to feel that my family origin was good—my father had joined Chairman Mao's revolution in the Jinggangshan period and had participated in the Long March. . . . Coming from this kind of 'meritorious' revolutionary family, what's wrong with getting a little privileged treatment?" Once Zhong arrived at the university, however, his conscience bothered him, and after a year of soul searching he decided to withdraw. "'Taking the back door' is using the power given to you by the people to serve your private interests," Zhong wrote. "In order to get your own children into the university, you don't go through mass recommendation, you don't go through the proper procedures of the party organization, but you use your own position and power, relying on interpersonal attachments and connections, to solve the problem. Some even use student recruitment slots as 'gifts' to hand out and take as they please, leaving the truly outstanding representatives of the workers, peasants, and soldiers outside the university gates. How can that kind of practice be serving the people?" Taking the backdoor, Zhong continued, not only was a betrayal of the public trust, but it indicated that cadres' children, such as himself, were becoming a privileged stratum. "It's not fair to use a phone call to get into the university. Particularly for a person who has come to regard himself as superior to others, who has always enjoyed privileges, this is an even more dangerous way of thinking. If this is allowed to develop, it would be easy for such people to become a 'privileged stratum,' to become revisionist. For a long time, cadres' kids have grown up in a favorable environment and have been divorced from the masses of workers and peasants and from productive labor. If they are not careful, it would be relatively easy for them to fall prey to bourgeois thinking."[10]

During the weeks following the publication of Zhong's letter of resignation, a series of commentaries and similar letters of self-criticism from children who had "taken the backdoor" and from officials who had "opened the backdoor" were published. Capitalist roaders in positions of power, it was argued, were encouraging such dishonest practices to corrupt other party officials and turn them down the road of revisionism. Thus, the practice was not simply a matter of individual wrongdoing, but a manifestation of class struggle. On January 25, 1974, Jiang Qing, Chi Qun, and other radical leaders launched the Criticize Lin Biao and Confucius movement by summoning party cadres to attend a "ten thousand person" meeting in Beijing. The new movement was an effort to rejuvenate the radical agenda of the Cultural Revolution, and they placed the Zhong Zhimin case and the campaign against taking the backdoor front and center at the meeting.[11] Several observers have suggested that the radical faction intended to use the campaign to once again mobilize workers and peasants to attack party officials, as they had in the early years of the Cultural Revolution. The radicals, Wang Fan wrote, hoped to "take advantage of the masses' dissatisfaction with 'taking the backdoor' . . . to launch an attack on a large number of central and local party, government, and military leading cadres."[12] Their aim, according to Zhang Libo, was to launch a "second Cultural Revolution" by linking the ideological messages of the Criticize Lin Biao and Confucius movement with a practical issue that could galvanize attacks on Communist officials.[13]

We will never know what direction the campaign against taking the backdoor might have taken or the impact it might have had, because it was scuttled less than five weeks after it began. Conservative leaders Zhou Enlai and Ye Jianying wrote letters to Mao expressing their concerns about the disruptive potential of the campaign and they won his concurrence. Responding to Ye's letter, Mao wrote that taking the backdoor was a huge problem that involved millions of people, but he agreed that Jiang Qing and Chi Qun's response was too extreme.[14] The radicals had just won Mao's support to launch the Criticize Lin Biao and Confucius movement, but he was not prepared to support another major assault on party officials. On February 20, the Central Committee decided, on Zhou's suggestion, that the campaign against taking the backdoor should be put on hold while the problem was studied and appropriate measures identified. The campaign was never revived.[15]

Mass Recommendation and Local Cadres' Power

By transferring responsibility for selecting college students to factories and rural production brigades, the recommendation system reinforced the already enormous importance of these work units in the lives of their members.

The work unit system established fixed collectivities that tied individuals to their workplaces, limiting mobility and, by the same token, fostering strong membership rights. Power in work units was concentrated in the hands of party cadres, who had a great deal of control over the lives of their subordinates, and the recommendation system gave even greater scope to their power. At the same time, because the power to recommend was supposed to be in the hands of rank-and-file workers and peasants, the system—at least theoretically—provided opportunities for democratic participation in decision making by work unit members.

In my investigation into how the recommendation system worked in practice, I was particularly interested in local cadres' power. My main sources were thirty-five people who told me about their direct experiences with the process. They had a variety of vantage points. Some were young people who were recommended by their villages or factories to attend Tsinghua. Others were urban middle school graduates who were sent down to live in villages or worked in factories during the Cultural Revolution decade and were eligible for recommendation, but were not recommended. Some had attended Tsinghua's attached middle school, while others were graduates of other middle schools who tested into Tsinghua University after examinations were restored in 1977. Still others were Tsinghua University teachers who were sent to various localities to recruit students.[16] Their accounts indicated that there was great variation in how recommendation policies were implemented. In some work units, the leading cadres simply picked the candidates themselves, while in others there were careful deliberations involving all members at the basic level and one or more rounds of voting. Their accounts provide only anecdotal evidence, but from these anecdotes much can be learned about the dynamics of the recommendation process.

ATTEMPTS TO CURB CADRE ABUSE FROM ABOVE AND BELOW

In order to stem backdoor admissions and other forms of corruption of the recommendation system, proponents worked to develop checks on cadre abuse from above and below. The main check from above was the teachers sent by universities. Teachers were not only supposed to assess the educational and political qualifications of the candidates, but also to interview each of them and visit their home villages and factories to investigate the processes through which they had been recommended. I interviewed two Tsinghua teachers, both of whom had originally been worker-peasant-soldier students, who traveled to distant provinces to recruit students. They took their task of investigating candidates very seriously and they both rejected well-connected candidates whose qualifications were not up to par.

Zhang Cuiying, who together with another teacher was sent to recruit new students for Tsinghua in 1974, spent over a month visiting villages and factories to interview candidates and other members of their work units. Zhang recalled that the extent of mass participation in the recommendation process differed greatly from place to place. "Whether or not it was democratic depended on the leaders of the work unit," she told me. "Some units held meetings where everybody made recommendations, and then the recommendations were sent up each level. In others they didn't have any meetings; the leaders just nominated people and the masses didn't know anything about it." She and her colleague, therefore, made sure to talk to ordinary villagers and workers. "If they were not satisfied, if they thought the person was no good, then that shows there was something wrong," she said. "If the masses didn't support the person, we didn't want him."[17]

Zhu Youxian, who was also sent to recruit students for Tsinghua shortly after he graduated in 1974, described an incident in which he and another Tsinghua teacher were pressured by a prefecture cadre to admit his son, whom they considered unqualified. "We decided his cultural level was too low and his political performance was also no good," he told me. Zhu and a colleague went to the youth's work unit to talk to his fellow workers. "We asked for their opinions, 'How was his performance, how was his labor, how was his thinking?'" The people they interviewed were not shy about criticizing the well-connected youth. "The masses' impression of him was not good," Zhu recalled. "Because his father was a cadre, he would always bully people."[18] The Tsinghua teachers rejected the candidate, to the great displeasure of the prefecture leader, and later Zhu checked with the university department involved to make sure the boy did not somehow find his way into the school. Ultimately, Zhu discovered he had been accepted at the Shandong Petroleum Institute, a result—he assumed—of his father's influence. "But," Zhu said proudly, "he didn't get into Tsinghua!"[19]

Potentially, the most important check on cadre abuse was scrutiny from below. Although the extent of participation in the recommendation process by rank-and-file peasants and workers was very uneven, it was clear from interviewees' accounts that many local cadres were concerned about being criticized by their subordinates for improprieties. In the wake of party officials' bitter experiences during the early years of the Cultural Revolution, when cadres were publicly humiliated and removed from office after rebels accused them of abusing their power, many were careful to avoid actions that could be seen as corrupt. Although most cadres regained their posts, the experience was not forgotten and many were keen to demonstrate their willingness to listen to the masses.

Zuo Chunshan, who was recommended by his village to attend university in 1974, described how the system of poor and lower-middle peasants' representatives, created during the Socialist Education movement and the Cultural Revolution to exercise mass supervision over cadres, impacted the recommendation process in his commune. These representatives were selected by the members of each production team to keep an eye on the cadres' handling of team, brigade, and commune affairs. (For instance, there were two locks on the door of the team's grain storage building, with one key in the hands of a team's accountant and the other in the hands of the peasants' representative.) When the commune was given a quota of students to recommend for university study, each brigade recommended candidates to the commune leadership, who then had to select the best candidates to send on to the county level. At that point, Zuo recalled, the leaders called a meeting of the peasants' representatives from all of the brigades in the commune. "The method of recommendation had to be discussed with them," he said. "Whatever the poor and lower-middle peasants said, it had to be done that way." In the end, commune leaders decided that the final list should be decided by means of an examination (a risky decision in the prevailing political climate). Zuo was convinced that fear of the peasants' representatives prevented commune leaders from abusing the recommendation system. "They wouldn't dare—there was poor and lower-middle peasants' supervision!" he told me. "Say a local cadre . . . used some maneuver to send his children or relatives to school. Maybe some of the poor and lower-middle peasants would think, 'How did he get to go? His qualifications are not as good as so-and-so's, maybe it's just because of his family relations.' So they would report it (*fanying*)."[20]

Many of those I interviewed believed the recommendation process in their own work units was aboveboard because it was difficult at that level for cadres' actions to escape scrutiny. For instance, Wu Xianjie, a middle school graduate who was not recommended by his factory, was convinced that those who were recommended were not the beneficiaries of cadre intervention. "If that happened at that level, the workers would be against the factory," he recalled. "Back then, the workers really had very high power. . . . If we had bad people, we could . . . go to a higher official to criticize the manager of the factory. . . . If they used some money [for themselves], if they recommend people not through the group and the workshop, they can be criticized." On the other hand, Wu—like others I interviewed—believed corruption was more of a problem at the county level because higher-level cadres were not subject to scrutiny from below. "In the factory . . . we have power; we can easily stand up," he told me. "[But] when you get to a higher level, our power is not there anymore."[21]

Typically, less than half the candidates who were recommended by their work units and sent on to the county level were actually selected to go to college. County officials could use their discretion in distributing quotas to lower levels and could use their considerable influence over subordinate officials to help determine names on the lists of candidates they received. Moreover, the criteria used to select among candidates at the county level often became a matter of contention and negotiation between county officials and university representatives, and there was room for individuals on both sides to manipulate the process on behalf of particular candidates.

RECOMMENDATION AND WORK UNIT POLITICS

Recommendation to go to college became a process of political deliberation similar to recruitment into the Youth League and the party. The political qualifications considered—in terms of both family origin and personal performance—were based on the same principles. As in league and party admissions, in addition to formal political criteria, candidates' relationships with leaders and other members of the work unit were critical to success. Unlike political admissions, however, which were decided by members of the league or the party branch, deliberations to recommend students were supposed to include all members of the work unit. And when the masses were actually involved, their participation changed the process in important ways because the perspectives of cadres and ordinary workers and peasants were often different.

The most important qualification—and the one most open to divergent evaluations—was a candidate's personal *biaoxian* (performance). As noted previously, this involved commitment to Communist ideology, support for the party, willingness to work hard, and a collectivist spirit. Many interviewees stressed that when work unit leaders considered a member's *biaoxian*, they were often looking for compliance with authority and personal loyalty. The power to recommend added to the incentives cadres could use to elicit compliant behavior among subordinates. The way this power could be used was illustrated by Cai Jianshe, a graduate of Tsinghua's attached middle school who was sent-down to live in a village during much of the Cultural Revolution decade. Born into an intellectual family in Beijing, Cai had always expected to go to college, but after examinations were replaced by recommendation, accomplishing this goal depended on the support of village and commune cadres. Cai, however, was inclined to resist authority, a trait cultivated as a rebel activist during the early years of the Cultural Revolution, and he soon crossed commune officials, closing the door to recommendation. He got into trouble after he was asked to serve on a team of commune members assigned to curb graft and corruption among cadres. Later, several members of the team were recommended to be village schoolteachers or factory workers. "They had curried favor with the cadres," Cai told me. "Those who didn't *pai mapi*—there were three or four of us—were sent back to the production brigade [to work in the fields]. The commune cadre in charge of the [urban] educated youth told me that I would never be recommended—I would be there for the rest of my life."[22]

When rank-and-file members of a work unit were involved in the recommendation process, they tended to place the greatest emphasis on an individual's work performance (*gongzuo biaoxian*) and less on political performance (*zhengzhi biaoxian*), and they typically had a contemptuous attitude toward *pai mapi* (literally, to "stroke the horse's ass") behavior. All of the people I interviewed who had been involved in basic-level discussions about recommendation, either in factory workshops or village production brigades, stressed the central importance of work performance in these deliberations.[23] Lu Baolan, who was admitted to Tsinghua in 1973 after she won recommendation by her village, insisted that villagers were most concerned about how hard the candidates worked. All of the production teams met separately, with every member attending, including those who were eligible for recommendation. Each team recommended two young people, both of whom could be members of any of the village's six teams. "The main thing was everybody's impression of you, every aspect of your *biaoxian*," she told me. "If there's anything you're not willing to do, then that doesn't make a good impression. You work hard, you're polite (*you limao*), you can endure hardship (*chiku*), when there's work to do, you're always out in front."[24]

Collectivist spirit was also highly evaluated by work unit members. Huang Jingshan said his fellow villagers recommended him because he worked hard on village projects, including irrigation works and a village library. He described a recommendation process very similar to the one in Lu's village, and said that he and a village girl were supported by the greatest number of the production teams in their brigade. "It was easy for villagers to identify people—they chose those who played an active role in the village," he told me. "We were both very active in village affairs—we volunteered to do difficult things, we had public spirit. I worked on irrigation projects—that was tough work. The villagers can tell how hard you work. . . . The standards were public spirit and contribution to the community."[25]

Both Huang and Lu stressed that villagers did not place a lot of weight on political rhetoric or Youth League activity. A candidate did not necessarily have to be the most active politically, Lu told me, but you had to "get along with people and help people." For ordinary villagers, Huang made clear, hard work and public spirit trumped political activism. "People did not choose the secretary of the village Youth League because he did not work very hard. He was involved in meetings too much, so he wouldn't work. He tried to get away from physical labor. . . . When you only say high-sounding words, if you were only politically active, people didn't like that."[26]

Feng Xiaobo, who was recommended by her factory workshop in 1976, echoed Huang's view, describing how political activism could actually generate antipathy among the rank and file. "If someone followed the leaders too much, the workers didn't like that person. To be an activist (*jijifenzi*) means you follow, you carry out the leaders' meaning and push it down. It's like being a boss. To be an activist, you have to be like that." Being politically active, however, did not foreclose having good relations with your fellow workers. "Of course, if you had a good relationship with people, you can also be active," she added. "If you do your own job well and also help other people, if you work a double shift, if your job skills are really good, people admire you. But if you're just politically active, but don't really work hard, they think you are politically active just to *pai mapi*, to please the leaders. If you say good things, but don't do good things, if you just talk empty rhetoric (*shuo konghua*), then they don't like you."[27]

It should be noted that Lu, Huang, and Feng not only worked very hard, but they were also very politically active. Lu was in charge of propaganda in her village, Huang was deputy secretary of the village Youth League, and Feng was a leader in her factory Youth League organization.[28] They each understood, however, that if political activism was not accompanied by hard work, it could hurt a person's standing among fellow work unit members. As Feng made clear, this was especially true if you seemed to simply be saying the right things to please the leadership and you were too eager to do their bidding. Their testimony reveals contradictory perspectives that quietly impacted the mass recommendation process. While catering to those in authority was a positive trait in the eyes of cadres, ordinary workers and peasants saw it as a negative trait. We can surmise that wherever the recommendation process was more democratic, it would discourage this kind of clientelist behavior. The inclusion of the rank-and-file workers and peasants in the process introduced a dynamic that worked against prevailing patterns of vertical dependence between cadres and their subordinates.

No matter how democratic the recommendation process might have become, however, it would always have accentuated the importance of *renqing* and *guanxi*, that is, personal relations and connections, respectively. While a more democratic process would have limited cadres' power over their subordinates, it also would have increased the importance of getting along with one's fellow work unit members. Friendship, factional affiliation, kinship, and other kinds of personal ties inevitably would have remained important factors. This could hardly be avoided in a system that made selection a political process.

Characteristics of Worker-Peasant-Soldier Students at Tsinghua

The replacement of examinations with work unit recommendation radically changed the social composition and qualifications of the Tsinghua student body. In 1970, university leaders produced a detailed report about the characteristics of the first cohort of worker-peasant-soldier students to enter the university (see Table 8.1). Unfortunately, such detailed reports are not available for subsequent cohorts. Below, I discuss the data provided about the 1970 cohort, adding suggestions about how the characteristics of subsequent cohorts changed, based on less detailed statistical data and interviews with students.

CLASS AND GENDER COMPOSITION

Recommendation altered the class origin of Tsinghua students dramatically. Before the Cultural Revolution, the great majority of students came from the old educated or the new political elites; in 1970, this was emphatically no longer the case. The proportion of students of "exploiting class" origin (ac-

TABLE 8.1
Tsinghua University worker-peasant-soldier students, 1970

Characteristic		Number	Percent of total
Total		2,842	100.0
Gender	Female	573	20.2
	Male	2,269	79.8
Class origin	Worker, poor or lower-middle peasant	2,304	81.0
	Upper-middle peasant, white collar	293	10.3
	Revolutionary cadre	105	3.7
	Exploiting	7	0.2
	Other	133	4.7
Work unit origin	Old worker	596	21.0
	Young worker	794	27.9
	Village youth	1,008	35.5
	Soldier	444	15.6
Education	Primary school	258	9.1
	Junior middle school	1,935	68.1
	Senior middle school	533	18.8
	Specialized middle school	109	3.8
	College	7	0.2
Political membership	Communist Party	1,431	50.4
	Communist Youth League	1,033	36.4
	None	378	13.3

SOURCE: Tsinghua University (1975). This report included students at Tsinghua's branch school in Sichuan Province.

cording to the CCP's taxonomy) fell from 10 percent to 0.2 percent, while the proportion from middling class origin (mainly urban educated groups, such as white-collar employees and independent professionals) fell from 46 percent to less than 15 percent (even counting all students from the "other" category among this group). While students who attended Tsinghua in the mid-1960s had estimated that children of revolutionary cadres made up about 9 percent of their classmates, such children made up only 3.7 percent of the 1970 cohort.

On the other hand, the proportion of students who came from the least educated classes, workers and poor and lower-middle peasants, grew from less than 37 percent before the Cultural Revolution to 81 percent in 1970.[29]

The proportion of students from both old educated and new political elite families undoubtedly increased in subsequent cohorts. In 1970, many of the urban "educated youth" who had been assigned to work in villages and factories after middle school had not yet completed the two years of manual labor required before becoming eligible for recommendation. Starting in 1972, quotas sent to rural communes regularly specified a number of slots reserved for sent-down urban youth, and many university recruiters favored these candidates, who were often better prepared academically. Although family origin became an even greater handicap for children from intellectual families during the Cultural Revolution, some were able to overcome this handicap by relying on hard work, educational advantages, and in some cases, on connections in the education establishment.

Revolutionary cadres' children had many of these same advantages and often had more influential backdoor connections, and there is no doubt that they were overrepresented among university students during this period (as they had been before the Cultural Revolution).[30] Moreover, revolutionary cadres' children were able to secure places in highly prestigious majors. In 1970, over half of the 105 revolutionary cadres' children admitted into Tsinghua were concentrated in two (out of eleven) departments—petrochemical engineering and electronics—and another 19 percent were in the university's branch school in Sichuan, which specialized in military-related electronics. This departmental concentration seems to have continued in later years, and the Petrochemical Ministry appears to have been particularly influential in placing cadres' children at the university.[31] Among the students recommended to study at Tsinghua during this period was Xi Jinping, the son of a top party official, who today is in line to succeed Hu Jintao as the party's general secretary.

Although it is certain that the proportion of university students from intellectual families and revolutionary cadre families continued to outstrip their tiny proportions of the general population and grew over time, they remained a small minority at Tsinghua throughout the late years of the Cultural Revolution. Interviewees who attended Tsinghua during this period reported that only a few of their classmates were from intellectual families, and most students in majors in which revolutionary cadres' children were not concentrated did not know a single student who was from a revolutionary cadre family.

The 1970 Tsinghua report provides little help in ascertaining the number of students who were children of lower-level cadres who did not qualify for the "revolutionary cadre" designation because their association with the CCP only began after 1949. The son of the head of a village production brigade who joined the party in the 1950s, for instance, was officially designated as being of peasant origin. Rural cadres, whose children typically lived in the commune where they held office, were in a particularly advantageous position to manipulate the recommendation system on behalf of their own offspring.[32] As a result, many of the rural origin students at Tsinghua and other universities during this period were children of rural cadres. Nevertheless, many others—including all of the rural origin students I happened to interview—came from ordinary peasant families.

Because the 1970 report did not differentiate between students of working-class and peasant origin, it is difficult to estimate the number of students of rural origin. The category "village youth" included urban young people sent down to villages after graduating from middle school, and while the great majority of soldiers were of peasant origin, not all of them were. On the other hand, many of the students recommended by factories were originally from peasant families. One interviewee, for instance, reported that among twenty-five classmates who were from factories, he and eighteen others had grown up in villages.[33] It is also difficult to use interview data to estimate the proportion of students who were of rural origin, because students' backgrounds varied tremendously by major. Several interviewees reported that the majority of their classmates had grown up in villages, while others said there were few peasants in their classes. Despite the lack of precise data, both documentary and interview sources support two broad conclusions: rural China continued to be greatly underrepresented at Tsinghua during the Cultural Revolution decade, but the proportion of students of peasant origin was higher than at any time before or since.

Part of the reason that the rural population was underrepresented was that Tsinghua was an industrial engineering school, and during this period students were in principle supposed to return to the work units that recommended them. Therefore, quotas for specific university departments were often sent to factories in related industrial branches. A large portion of chemical engineering quotas, for example, were filled by workers from chemical refineries, while many metallurgy majors were sent from steel mills and foundries. For this reason, students of working-class origin were greatly overrepresented at the university.

Although the available data leave important questions unanswered, it is clear that the recommendation system radically changed the class composition of the Tsinghua student body. Before the Cultural Revolution, children of the old educated elite and the new political elite made up about two-thirds of Tsinghua students. Now, even though it is likely that both groups—and especially the latter—continued to be overrepresented, the campus was filled mainly with children of workers and peasants. The recommendation system had significantly redistributed opportunities to attend college in favor of the less educated classes.[34]

The gender imbalance at Tsinghua, in contrast, changed little under the recommendation system. In 1970, only a little over one-fifth of all students were female, about the same proportion as before (and after) the Cultural Revolution decade.[35] There was, however, a significant change in the gender composition of rural origin students. Before the Cultural Revolution, a significant portion of Tsinghua students came from the countryside, but they were almost all male. Very few village girls made it into senior middle school and even fewer competed in the college entrance examinations. The same situation prevails today. Among the fourteen students I interviewed who grew up in villages and tested into Tsinghua before and after the Cultural Revolution, not one was female. Moreover, in my detailed inquiries into the social composition of interviewees' classes before and after the Cultural Revolution, no student reported having a female classmate from the countryside. This reality resulted in an imbalance that several romance-conscious students noted: while there were quite a few rural boys at the university, there were only urban girls in attendance (presenting a social gap difficult to bridge even during the Mao era). In contrast, among interviewees who attended Tsinghua in the early 1970s, many recalled village girls in their classes. Indeed, three of the eleven worker-peasant-soldier students I interviewed were women who grew up in villages.[36]

EDUCATIONAL AND POLITICAL QUALIFICATIONS

The 1970 report documented the low and uneven educational level of the first cohort of worker-peasant-soldier students at Tsinghua. Over 9 percent had only been to primary school, 68 percent had only junior middle school education, and less than 19 percent had been to senior middle school. Moreover, many had attended poor-quality schools in villages and working-class districts and all had seen their education interrupted by the factional fighting of the early Cultural Revolution years. Nevertheless, they were a well-educated group compared to the general population. In 1970, only about 10 percent of the eligible age group had graduated from junior middle school and only about 3 percent had graduated from senior middle school. At that time, education officials could not have set graduation from senior middle school as a standard for recommendation without eliminating the vast majority of workers' and peasants' children from the pool of eligible college students. Over the course of the Cultural Revolution decade, the annual number of senior middle school graduates increased from less than 300,000 to well over 6,000,000. If the rapid expansion of middle schools had continued, the Cultural Revolution goal of making senior middle school education universal might have been accomplished relatively soon, providing a more uniform educational base among the young people eligible for recommendation.[37]

Even if senior middle school had become universal, however, worker-peasant-soldier students never would have been characterized by the uniformly high standard of academic preparation that selection by examination guaranteed. For one thing, the length of primary and secondary schooling had been cut, middle school education during this period did not prepare students for more complex college-level instruction, and the system of elite college preparatory schools had been eliminated. Moreover, the recommendation system was primarily designed to select for political characteristics rather than academic proficiency. Members of a candidate's work units were not in a position to really evaluate his or her academic abilities. In their deliberations, work unit members often gave preference to candidates who had been to senior middle school, and they considered intelligence and abilities demonstrated on the job, but they knew little about how they did in school. "People didn't really consider education," recalled Lu Baolan, who was recommended by members of her village to study at Tsinghua. "They didn't really know if you studied well or not." Moreover, the prevailing thinking did not place much stress on performance in school. Wu Xianjie, a youth from an educated family who worked in a factory during the Cultural Revolution decade, recalled that when his fellow workers discussed candidates for university recommendation they were not very concerned about their academic preparation. "They thought, when you go to school, everything will be changed—even if you don't know anything, you can learn; they thought, 'He's good, he's a good worker . . . he'll be a good student anyway.'"[38]

The teachers sent by universities to recruit students ultimately had the main responsibility for evaluating candidates' educational preparation and abilities, and they were also more concerned about finding recruits who were better prepared for university study. Nevertheless, these teachers were also instructed to review both political and cultural qualifications, and not simply look for the candidates who were strongest academically. Teachers administered tests to evaluate candidates' educational background and learning ability, but the tests were typically ad hoc and designed to assess whether candidates met minimum expectations, rather than to identify the most proficient. For instance, Zhu Youxian, a Tsinghua instructor who served as part of a team sent to recruit students in 1974, told me that they gave all the candidates a written test designed by local education officials, but only used the scores as a reference, not as the deciding factor in whom to recruit. They gave priority to political qualifications, and were flexible when evaluating candidates' educational qualifications. "We would talk to the candidates to see their ability to respond, to see how their math was, to see what their education level was, if they were junior middle school or senior middle school graduates, to see how smart they were," he told me. "Their education level could not be too low. Still, one might be a senior middle school graduate but not so smart, while another might only be a junior middle school graduate but he might be smarter."[39]

Although the academic preparation of worker-peasant-soldier students was relatively low, politically they were a highly select group: according to the 1970 report, over half of the new students were already party members and most of the rest were Youth League members. The Tsinghua newspaper also proudly noted that 268 of the first cohort of 2,842 worker-peasant-soldier students had been activists selected to participate in county conferences to study Mao Zedong Thought, and 351 had served on local revolutionary committees as representatives of mass organizations (the factions that emerged during the early years of the Cultural Revolution).

The recommendation system was designed to select young activists who were hardworking, capable, cooperative, and politically loyal, and who were ambitious, but willing to subordinate personal interests to "serve the people." Zhang Cuiying, a member of the 1970 Tsinghua cohort, was this type of activist. Today she still displays the enthusiasm and confidence she had as a young activist and her recollections of those days are brimming with the Communist slogans of the period. Her parents were ordinary villagers, who had been designated part of the poor and lower-middle peasants during Land Reform. In the mid-1960s, she became the second person in her middle school class to join the Youth League and she was active in the student movement during the Cultural Revolution. After returning to her village, she worked in the fields and sometimes as a substitute teacher in the village primary school. She was chosen to head the village women's association, and in this capacity, she visited Dazhai, the model production brigade in Shanxi Province. After she returned, she organized an "iron girls' breakthrough team" (*tie guniang tujidui*) in her village. The team, the only one in her area, was composed of ten young women who volunteered to do especially heavy tasks, such as moving earth and water. "We girls were not weak," she said proudly, "we could do everything the boys could do."[40] In 1970, Zhang represented her village in a county conference to study Mao Zedong Thought.

The worker-peasant-soldier students who attended Tsinghua during the 1970s were, by all accounts, a hardworking group. They studied very diligently, getting up early in the morning and staying up late at night. Although teachers lamented their lack of educational preparation, they praised their enthusiasm and determination. "Most worker-peasant-soldier students studied very hard," said Liang Yousheng, who taught at Tsinghua from 1958 until a few years ago. "I would say they studied harder than students do today."[41]

Mass Recommendation and the Reproduction of Educated and Political Elites

The recommendation system contained within it the essential elements of the Cultural Revolution class-leveling strategy. On the one hand, it was intended to redistribute access to education and to make selection a political process based largely on political qualifications. On the other hand, it was designed to promote wide participation in political deliberations. The goal was to first transfer power from the cultural to the political field, and then disperse political power. The system was more successful in the first endeavor than in the second.

The main purpose of mass recommendation was to check the social reproduction of the educated elite. The examination-oriented school system had been a powerful mechanism that reproduced the skewed distribution of cultural capital, and replacement of examinations by recommendation was intended to facilitate a radical redistribution of educational opportunities. It was very effective in accomplishing this goal. First, the advantages enjoyed by children from educated families under the examination system were severely undermined and the demographic dominance of these children in the university system was greatly diminished. Second, elimination of the examinations facilitated a massive expansion of the education system and allowed schools to focus on providing basic education for the many rather than on selecting a few for higher training.[42] As a result, recommendation severely disrupted the transmission of cultural capital across generations and facilitated a significant dispersion of educational attainment.

At the same time, the recommendation system had the potential to become a powerful mechanism of class differentiation based on political capital. By making selection for college a political process based largely on political criteria, recommendation expanded the scope of political credentialing at the expense of academic credentialing. The recommendation system could be used to reinforce clientelist relations in work units, as cadres used their power to recommend to encourage greater personal loyalty among subordinates. Moreover, backdoor admissions became a means to reproduce the political elite by facilitating the conversion of parents' political capital into children's cultural capital.

The radical advocates of recommendation—who were implacable enemies of the party officialdom—attempted to prevent the system from becoming an instrument for the reproduction of political capital, and they did this principally by trying to make the system more democratic. By institutionalizing mass participation in recommendation, radical advocates hoped to redistribute political power in work units, enhancing the power of the rank and file at the expense of cadres. Their fledgling efforts to mitigate problems engendered by the concentration of power in the hands of cadres by promoting rank-and-file participation and launching a campaign against backdoor admissions were intriguing and bear further study. Interview testimony indicated that mass participation in decision making did, indeed, limit cadre abuse and reined in the recommendation system's tendency to promote clientelist behavior. Nevertheless, these problems plagued the system throughout its brief existence, and the results of mitigation efforts do not inspire confidence that proponents of recommendation had found effective means of preventing the system from facilitating class differentiation based on political capital.

PART FOUR

The New Era (1976–Present)

Chapter Nine

Rebuilding the Foundations of Political and Cultural Power

On October 6, 1976, less than a month after Mao Zedong's death, troops dispatched by the Beijing garrison of the People's Liberation Army took control of Tsinghua University and arrested Chi Qun and Xie Jingyi. At the same moment, Zhang Chunqiao, Jiang Qing, Wang Hongwen, and Yao Wenyuan, who would thereafter be known collectively as the Gang of Four, and other important radical leaders were arrested and military units took control of key mass media facilities and government offices.[1] Thousands of others were subsequently arrested around the country as the radical faction in the CCP was systematically suppressed. Thus, the elaborate regime of institutionalized contention between rebels and administrators that Mao had constructed during the previous decade was dismantled within weeks of his death.

The Communist era in China is conventionally divided into two periods bisected by Mao's death. There was much continuity between the two periods, as the CCP retained its monopoly on power and continued to pursue a vision of modernizing China. The main difference was that after 1976 the party abandoned its program of eliminating class distinctions. Deng Xiaoping, who led the regime that emerged out of the power struggle that followed Mao's death, ushered in a New Era in which Mao's insistence on perpetuating class struggle was emphatically repudiated. Deng declared that China could only achieve national prosperity if it allowed some people to get rich first, with the clear implication that class differences were not only unavoidable, but also desirable.[2] Suddenly, the convulsive tensions that class leveling had caused over the past three decades were relieved. During the first years of the New Era, a technocratic class order was consolidated in China, based largely on possession of political and cultural capital. This order depended on the academic and political credentialing systems that have been central subjects of this book. These institutions had played a critical role in class differentiation throughout the Communist era, but they had always been constrained by the egalitarianism of the Communist program, and they had been systematically undermined by the class-leveling policies of the Cultural Revolution decade. Now they could be rebuilt and enhanced in an orderly fashion, freed from the constraints, uncertainties, and destructive mobilizing power of class-leveling ideology. Under new leadership, Tsinghua quickly restored the elite educational machinery and monolithic party organization of the Jiang Nanxiang era, and as the political and academic credentialing systems were reconstructed nationwide, the university once again established its place at the top of both. This chapter will look in detail at the evolution of each of these systems during the post-Mao era.

Rebuilding the Political Foundations of Class Power

After a new leadership team arrived at Tsinghua in the spring of 1977, the university was subjected to a thoroughgoing Great Clean-up and Investigation campaign. Mass rallies were held to denounce Chi Qun and Xie Jingyi as well as several faculty members who had become known around the country for writing radical treatises. Kuai Dafu and other radical figures from the early years of the Cultural Revolution were also brought back to Tsinghua to be publicly criticized before being sent to prison. The workers' propaganda team was dismissed and its members were sent back to their factories and military units. Investigations were conducted into the political records and inclinations of everyone who had served in a leadership capacity at any level under the propaganda team. Many were removed, and those whose political reliability was dubious were required to take part in political study classes to "unify thinking" (*tongyi sixiang*).[3]

In 1978, a little more than a year after the Great Clean-up campaign had begun, Deng Xiaoping assigned Jiang Nanxiang to head a team to investigate the situation at the university. In his report, Jiang made it clear that no expression of support for the now discredited policies of the Cultural Revolution decade would be tolerated, calling attention to numerous incidents that betrayed remnants of "old thinking." He pointed out by name teachers, workers, and worker-peasant-soldier students who had expressed discontent with the new leadership and its policies, drawing particular attention to the intransigence of some of the university's workers. "Under the poisonous and corrosive influence of Chi and Xie, some workers fancied themselves to be leaders and reformers, and they became the vanguard of rectifying intellectuals," Jiang wrote. "We have to be aware that even now some workers can't change their way of thinking and can't accept the line of the party's Eleventh Congress."[4]

In the end, the campaign completely eliminated any overt manifestation of the factional contention that had gripped the university for over a decade. Jiang Nanxiang was appointed minister of higher education, and a longtime associate, Liu Da, was selected to take over as party secretary at Tsinghua. Like Jiang, Liu had been active in the anti-Japanese student movement in 1935 and the two subsequently worked together in the leadership of the Youth League; before the Cultural Revolution, Liu had headed the Chinese Technology University. In 1977, he brought several top Youth League officials to Tsinghua with him to help straighten out the university and purge it of radical influence, but he ended up largely relying on He Dongchang, Ai Zhisheng, and other pre–Cultural Revolution Tsinghua-brand cadres to reorganize and run the school.

With the removal of the workers' propaganda team, the new university leadership reconstructed the party and administrative hierarchies in line with the ideals of bureaucratic efficiency and monolithic unity. No longer obliged to contend with radical opponents and outside interlopers, they rebuilt a single chain of command. The system of administrative ranks was dusted off and refurbished, removing ambiguities, promoting long-deserving cadres, and eliminating the abnormal condition of higher-ranking cadres reporting to lower-ranking cadres.

RELIEVING THE MASSES OF THE BURDEN OF POLITICAL PARTICIPATION

Under the banner of political unity, the divisive practices and institutional arrangements that Mao had promoted during the Cultural Revolution were eliminated. "Speaking out freely, airing one's views fully, writing big character posters and holding big debates," declared Deng in 1980, "taken as a whole . . . have never played a positive role."[5] These practices had been formally protected since the radical faction succeeded in writing them into the Constitution at the Fourth National People's Congress in 1975. At Deng's urging, all of these rights, together with the right to strike, were rescinded by the Fifth National People's Congress in 1980.[6] Students "attending, administering, and reforming" the university and other practices associated with the Cultural Revolution were now dismissed as "old thinking" and a hindrance to the development of "scientific management," which became a catchword of the New Era. After the decisive victory over the radicals put an end to the deep factional divisions in the party, party leaders strived to resolve their differences through orderly procedures and behind closed doors; there was no longer any reason to involve the masses in political disputes.

Politics, which had pervaded virtually every aspect of society, now receded from the lives of ordinary people. During the Mao era, individuals had been expected to make a commitment to the Communist project and to demonstrate this commitment in their behavior, and during the Cultural Revolution decade, politics was further animated by volatile factional conflicts that extended down to the base of society. Even after autonomous mass organizations were suppressed in 1968, institutionalized factional contention had kept tensions high. At Tsinghua, the control of the workers' propaganda team was uncertain and the campus was rife with contentious—and hazardous—political discussion and activity. Moreover, it was hard to avoid: students, workers, teachers, cadres, and members of the propaganda team all faced exacting expectations to participate politically. In the post-Mao era, in contrast, individuals were urged to attend to their own affairs and leave politics to party officials. "Extracting more oil is the politics of the petroleum industry," declared Deng in 1979, "producing more coal is the politics of coal miners, growing more grain is the politics of peasants, defending the frontiers is the politics of soldiers, and working hard in study is the politics of students."[7]

On the one hand, there was no longer room for public political contention or criticism of party officials; on the other hand, the depoliticization of private life allowed individuals much greater freedom to hold their own opinions. Literature and art, education, and science and technology no longer had to serve politics, experts no longer had to be Red, and individuals were no longer obliged to attend political meetings. As a result of the suppression of factional conflict and the general retreat of politics from day-to-day life, the political tension that had gripped people's lives during the Cultural Revolution decade dissipated, a development welcomed by many. In schools, factories, and villages across China, often onerous political obligations—long meetings, obligatory study of newspaper articles and party directives, daily discussion of work unit problems and plans, mandatory criticism and self-criticism, and unremitting political movements—were fading into the past. The receding of politics, however, was a mixed blessing. In recounting the political history of his rural county, Dongping Han lamented the fact that the dismantling of collective agriculture also led to the disappearance of a "public arena" for discussing village affairs.[8] This arena also disappeared in factories and schools. While peasants, workers, teachers, and students were freed from their political obligations, they also lost the opportunities that had come with these obligations to participate in decision making and hold their leaders accountable. As politics retreated behind party committee doors, political power in schools, factories, and villages was concentrated to an even greater extent in the hands of the local party secretary and other senior officials.

Fang Xueying, a former worker-peasant-soldier student who teaches at Tsinghua today, compared the highly politicized mind-set of students of her generation with the more individual concerns of her students today. "Students back then and students today are very different," she told me. "We always thought about the future of the country, because we thought our own future was intimately connected with the future of the country. Today students just think about their own future, it's all about individual struggle and individual planning (*geren fendou ziwo sheji*), they think about going overseas, about how they should develop themselves for their own future. Then we thought about the country's affairs and we were very concerned about reforming the school—our responsibility was to attend, administer, and reform. Today students don't have this responsibility—their responsibility is to study."[9]

COLLAPSE OF COMMUNIST IDEOLOGY AND THE COLLECTIVIST ETHIC

The retreat of politics was accompanied by the collapse of Communist ideology and the collectivist ethic. This collapse was quite sudden and was recorded dramatically in my interviews. While students who attended Tsinghua between 1949 and 1976 often used collective goals to describe their own motivations and those of their colleagues, these goals were largely absent from the vocabularies of students who attended Tsinghua after 1977. Those who grew up during the Mao Zedong era consistently invoked a sharp contrast between the past and the present in terms of beliefs and ethical dispositions. Speaking from today's perspective, they often described their past collectivist ideals as naive, but they continued to remember those ideals with a sense of nostalgia and loss. Although they clearly had not been indifferent to their own well-being and status, during the Mao era their personal ambitions were tied to collective organizations and were framed in terms of the wider Communist project. Subsequent generations, they said, could not understand the collectivist thinking that prevailed in the past, because younger people had only known an orientation toward individual pursuits.

In a set of poignant essays, Jiwei Ci observed that in the post-Mao era people experienced a loss of meaning as their vision of a Communist future collapsed. During the Mao era, he wrote, people had been urged to work hard and forgo immediate rewards in order to achieve a future of abundance; moreover, they were asked to struggle and sacrifice not for individual prosperity, or even for the prosperity of their families, but rather for collective prosperity. The sudden unraveling of this collectivist vision set in motion a rapid movement "from utopianism to nihilism to hedonism."[10] Ci's observations echoed those of dissident writer Liu Binyan, who described a "spiritual malaise" in the 1980s caused by the collapse of the collectivist ideology of the Mao era. "The popular ideological trends of the 1980s are diametrically opposed to the extremist ideology under Mao, having moved to the other extreme," Liu noted. "Back then, politics was everything; now it is nothing. Back then 'serve the people' was the essential revolutionary slogan; now disdaining the people is considered progressive. Back then, to advocate 'work only for the people, not for oneself,' was the norm, even if those who shouted it the loudest actually did the opposite; now it has become fashionable to advocate 'work only for oneself, not for the people.' Back then, intellectuals were to go among the masses; now they are permitted to live comfortably and in good conscience in their own closed circles."[11]

The Cultural Revolution did great damage to the perceptions that sustained a collectivist ethic by focusing attention on corruption and abuse of power by Communist cadres. Nevertheless, as we have seen in the testimonies and actions of Tsinghua teachers and students, the collectivist ethic retained its power during Mao's last years. It was still possible to call on people to sacrifice for the common good, while also asking them to do battle against transgressors of the collective trust, even though exposure of such betrayals aroused suspicions about the viability of the project. The general collapse of public confidence in collectivist ideals came only after the post-Mao leadership renounced these ideals. They did so explicitly and vehemently, denouncing the dangers of "egalitarianism" and the idea of "eating from one big pot," and instead they promoted the pursuit of individual success, whether through the examination competition or peasant entrepreneurship. Communist efforts to sustain collectivism were decisively abandoned in favor of inspiring hard work through individual rewards, and Communist asceticism was replaced with a new morality in which it was glorious to get rich. Despite the fervency it had once inspired, the collectivist ethic ultimately proved to be quite fragile. It had depended on the goal of a classless society; once this goal was abandoned, collectivism could not but collapse, for if some were to reap the rewards of others' hard work and sacrifice, the latter would not be heroes, but fools.

Having jettisoned the Communist class-leveling program, the new leadership redefined the party's mission exclusively in terms of economic development. After the heart of Communist doctrine—collectivism and a classless future—had been abandoned, however, the entire ideological system was largely shorn of meaning. Other ideologies also offered promises of modernization and economic prosperity. Communist claims in these spheres not only lacked novelty, they also lacked luster next to the accomplishments of wealthy capitalist countries. As a result, visions of a bright Communist future were increasingly replaced by more concrete images of contemporary success in the United States, Japan, and elsewhere.

Students who attended Tsinghua in the 1980s could scarcely be expected to commit themselves to an ideology so out of alignment with the actual practice—and much of the contemporary rhetoric—of the party. As a result, ideological requirements, which had previously been a critical part of Youth League and party recruitment process, were now reduced to largely meaningless formalities. Liu Wenqing, who joined the party while studying at Tsinghua in the early 1980s, recalled how difficult it was to write the essays required as part of one's party application: "It was so hard to find a reason to say why you believed in Communism, it was a big struggle," she told me. "The teachers knew it was hard, it was hard for everyone to find a reason, so we came up with an argument—that we all know society will grow better and better, and we all know Communism is the best, and so it will evolve into Communism."[12]

Liu's predicament illustrates the ideological sea change that had taken place in just a few years. Students who had attended Tsinghua during the Mao era (whether before or during the Cultural Revolution decade) might have found writing party application essays tedious, but few would have had any difficulty declaring their belief in the Communist project. During the heated factional battles of the Cultural Revolution, students on both sides were ready to die to defend their vision of a Communist future. By the time Liu arrived at Tsinghua in 1981, however, everything had changed. She and other students continued to strive to join the Youth League and the party, but they mainly regarded the effort as a means to advance their careers, and their motivations had little to do with commitment to Marxist ideology. Moreover, although political activism still involved attending meetings, with the depoliticization of society and the collapse of Communist ideology, the intensity of the commitment associated with Youth League and party membership also faded.

The changing attitudes among students at Tsinghua University and its attached middle school were deeply disturbing to veteran teachers and school officials. Their distress was recorded in an official school history written in the mid-1980s by leaders of the middle school. By then, the school had recovered from the disruptions of the Cultural Revolution. The former principal, Wan Bangru, was once again in charge, the disruptive students from nearby villages had been removed (along with all students from these villages), order had been restored in the classrooms, and the school had regained its keypoint status, allowing it to recruit from among Beijing's top-scoring students. But it was among these students—in whom Wan and veteran teachers had placed great hopes—that they confronted a phenomenon that was in some ways even more disturbing than the disruptions caused by village students. "The confused thinking of the students has reached a shocking state!" the school leaders lamented. "There are problems with student thinking even among those in the 'recruit the best' (*zeyou luqu*) program in the keypoint classes, and even among Youth League members and student cadres. They are averse to political study; they think it is 'fake, loud, and empty.' [They say], 'Only getting into college has meaning, everything else is fake.'"[13] Students in the 1980s, Wan and his colleagues believed, thought too little about the collective and too much about their own narrow concerns. The teachers' disappointment in the younger generation stood in contrast to their praise for students before the Cultural Revolution, who, they wrote, were not only bright and hardworking, but had lofty political ideals and were committed to building up the school and the country. The vitality of Tsinghua's attached middle school in the early

1960s was, in the history recounted by Wan and his colleagues, shattered by the Cultural Revolution and had not recovered since, especially in terms of morality.

This account of the decline of public spirit among elite middle school students echoes a more general narrative that reflects official discourse and is widely shared among many members of the generations who lived through the Mao era. In this narrative, before the Cultural Revolution, officials were honest; workers, peasants, and students were diligent and hardworking; and there was economic development and social order. This golden age was destroyed by the Cultural Revolution, which undermined people's trust in officials, disrupted economic progress, fractured social order and harmony, and opened the door to moral decline. After the Cultural Revolution decade, order was restored, but moral decline only grew worse. In another—less official—narrative, during the upheaval of the Cultural Revolution, people's eyes were opened to the seamy side of the Communist Party and they began to lose faith in its officials. In both narratives, a trust that existed between the people and their leaders was broken by the Cultural Revolution and was never restored. Regardless of whether individuals subscribe to one version or the other, or have yet another interpretation of this history, they express deep disappointment in what they see as the steep moral decline of Communist cadres and of the population as a whole. Moreover, rising cadre corruption in the post-Mao era is indelibly associated with the new motto, "to get rich is glorious."

The strong ethic of public service that had guided Chinese officials during the first decades of Communist power was ideologically linked to Communist ideals of collectivism, egalitarianism, and asceticism. The erosion of this ideological foundation severely undermined party cadres' ethic of public service. The rise of official corruption during the post-Mao era only further weakened the prestige and authority of the party, which had already been severely eroded during the Cultural Revolution. Although factional strife had been eliminated and a unified party hierarchy was reestablished after Mao's death, the damage done to the authority of party offices during the Cultural Revolution could not be completely repaired and Communist cadres never regained the type of authority they had previously enjoyed. The men who served as party secretary at Tsinghua in the 1980s presided over a hierarchy that was far more united and bureaucratically efficient than the fractured organization run by Chi Qun and Xie Jingyi during the late years of the Cultural Revolution, but none wielded anything like the kind of authority that Jiang Nanxiang had enjoyed before the

Cultural Revolution. The difference was not simply a matter of Jiang's abilities and character; rather the *office* of party secretary no longer commanded the kind of uncritical obedience that it had in Jiang's day. The criticism and humiliation to which party officials had been subjected during the Cultural Revolution, together with the denunciations of bureaucratic authority and political tutelage, had introduced a skepticism that made a return to the old order impossible.

RESTORING AND REFINING THE POLITICAL CREDENTIALING SYSTEM

The political credentialing system, which the Cultural Revolution had thrown into disarray, was rebuilt after Mao died. At Tsinghua, the workers' propaganda team had eliminated both the political counselors system and the teaching and research group that conducted political courses, and it had abandoned much of the rest of the apparatus—organized around the Youth League—that had specialized in recruitment and political education, denouncing it as a vestige of the "revisionist" party machine of the old Tsinghua. The new leadership resurrected the corps of cadres dedicated to political thinking work and it once again became the political heart of the university party organization. The political instructors' teaching and research group was rehabilitated and a "Department of Student Work" was established to oversee the operations of a reinvigorated Youth League.[14] Jiang Nanxiang's program of hiring select students and young teachers as political counselors, who coordinated political education and recruitment for several classes of students, was resurrected and received Deng Xiaoping's endorsement. "After several years of training, [political counselors] become a 'Red and expert' force for carrying out political work," Deng declared. "This was a good example."[15] Subsequently, universities around the country created political counselor systems based on Tsinghua's model.

The political credentialing system was rationalized in a manner that had been impossible during the Mao Zedong era. In the past, recruitment, promotion, and demotion had all been carried out in the course of political movements. Activists had been recruited in large numbers during political movements and relatively few were recruited during intervals of calm.[16] The fate of party leaders had been determined by rounds of mass criticism and shifts in factional power. Now that the turbulence of political movements was consigned to the past, recruitment, promotion, and demotion could proceed in a much more orderly fashion. The party's Organization Department was finally able to establish stable *nomenklatura* lists and orderly career paths, and the Central Party School began to play a greatly enhanced role, selecting and training upwardly mobile cadres.[17] Advancement up the party and state hierarchy could now be governed by more systematic selection procedures based on formal sets of criteria and credentials.

Shorn of its ideological meaning, party membership retained its instrumental value as a political credential and networking tool, attracting ambitious university students who aspired to public service and leadership positions.[18] As in the past, the party sought young people with leadership qualities, but the New Era brought significant changes in the definition of leadership. During the Cultural Revolution decade, the definition of leadership had included "rebel spirit." The freewheeling factional contention of the early years of the Cultural Revolution had made an attitude of defiance a worthy personal quality, and the radical faction had attempted to make rebel spirit into an important criterion for political recruitment in the early 1970s. At Tsinghua, although the leaders of the workers' propaganda team did not take kindly to criticisms of themselves, they had encouraged students and workers to criticize their immediate superiors. Now the propaganda team was gone, and a single hierarchy of authority had been restored. As a result, rebel spirit was discredited as a vice of the Cultural Revolution; students were expected to obey their teachers and workers were expected to obey their supervisors.

It was now easier for party functionaries, most of whom had never had much use for the rhetoric about rebel spirit, to select for leadership qualities required by a bureaucratic organization, seeking out young people who were able to work effectively in the party hierarchy, taking guidance from above and giving guidance to those under their supervision. There was no longer any place in the party for the discourse of class struggle; instead, the party now completely embraced technocratic values, celebrating pragmatism, organizational efficiency, scientific management, and political order.

The Tsinghua party organization quickly recovered from the taint associated with the radical leaders of the workers' propaganda team and regained its reputation as a premier training ground of political cadres. Indeed, as we will see in Chapter 10, after the new CCP leadership embraced a technocratic agenda, young cadres who had been cultivated by Tsinghua's party organization had credentials that facilitated rapid upward mobility in the party and state hierarchies.

Rebuilding the Cultural Foundations of Class Power

In May 1977, Deng Xiaoping opened the New Era with a call to "respect knowledge and talent" (*zunzhong zhishi, zunzhong rencai*).[19] The new CCP leadership not only repudiated the radical education policies of the Cultural Revolution, but completely disavowed class leveling in the cultural field, freeing the party from a doctrinal commitment that had complicated its education policies since 1949. The party was now able—for the first time—to build elite schools and embrace meritocratic ideals without moral ambivalence or political second-guessing. Under Deng's leadership, the government quickly rebuilt the academic credentialing system, explicitly enhancing and perfecting the school system's function as a mechanism of social selection.

RESTORING THE EXAMINATIONS

The first project Deng Xiaoping took on after he was rehabilitated in 1977 was restoring the college entrance examinations. The move was highly controversial and Deng, who had been charged with overseeing education, science, and technology for the State Council, only succeeded after forty days of sharp debate at a national meeting on university admissions held in the summer of 1977.[20] The debate continued after a new cohort of students (who had passed the examinations) joined the older cohorts of worker-peasant-soldier students on university campuses across the country. Cai Jianshe, the son of a highly educated intellectual who graduated from Tsinghua's attached middle school in 1968 and tested into Tsinghua University in 1977, described the arguments that ensued. "The relations between us and the worker-peasant-soldier students were very tense," Cai told me. "We thought we had the right to be there because we had tested in, and they thought they had the right to be there because they had the political and class qualifications."[21]

Cai and the other new students were rightly proud of their accomplishment. More than ten million candidates participated in the first round of college entrance examinations in December 1977, and less than 3 percent of them were admitted to college. Like Cai, many of the successful candidates had gone to keypoint middle schools before the Cultural Revolution, when curricula had been tailored to the examinations. Of these students, a great many were children of intellectuals and government officials, the two groups that had filled the classrooms of these schools.[22] Those who had passed the entrance examinations argued that exams were necessary to assure high standards, but they also believed they were a fairer method to select university students than recommendation. Li Huan, who like Cai had attended Tsinghua's attached middle school before the Cultural Revolution and passed the university examinations after they were restored in 1977, reiterated this argument. She had been deprived of the opportunity to go to college under the recommendation system, she said, because her grandfather had been a landlord and a Nationalist military officer. "A peasant kid might say the exam system is unfair to him, but in fact everyone has to depend on their own ability, on their own level," she told me. "Recommending a worker-peasant-soldier student was not like that—it was not based on your individual ability, right? Say he only attended primary school and doesn't know anything, but because his family origin was good, he could go to college, while those like us didn't have a chance." Exams are the fairest method, she insisted, because they are based on individual ability. "It doesn't matter if you are from a worker's family or a peasant's family—if you do well on the test, you can get in."[23] This argument carried the day and is now almost universally accepted in China.

In the late 1970s, university authorities at Tsinghua and elsewhere warmly welcomed the new students. The authors of a semiofficial history of Tsinghua and Beijing universities used colorful language to describe the changing of the guard. "After [the new students] tested into the university, they became the favorite pets and delicate children of the party and the state," they wrote. "[In contrast,] the worker-peasant-soldier students were in a sorry situation. They were older and their knowledge base was weak. Moreover, in school their purpose was not to study but to take the power to run the school away from the bourgeois academic authorities and help the old intellectuals remold themselves. They had studied herding sheep and building dams. When political power changed, they went from being doted on to receiving a cold shoulder. The previously submissive stinky intellectuals were suddenly wearing Western clothes, leather shoes, and gold rimmed glasses, and they adopted arrogant and scornful attitudes, carried thick bundles of teaching materials under their arms, and walked to class in a self-assured and haughty manner. The new students, who had come in by passing the exams . . . had a cocky attitude. The two groups [the new students and the worker-peasant-soldier students] rarely spoke; they seemed like feuding clans."[24]

Teachers and cadres at Tsinghua differentiated between the "real university students" (who had passed the examinations) and the worker-peasant-soldier students. The inferior status of the latter was reinforced by decisions to give them distinct diplomas and record their numbers separately in university records. Two hundred former worker-peasant-soldier students who had been retained to work at the university were transferred to other work units, and most of those who stayed were shifted to administrative work. Only a few, mainly those who tested into new postgraduate programs, were allowed to return to academic positions.

REBUILDING THE EDUCATION PYRAMID

Under the revived system of keypoint schools, Tsinghua and several other national universities received far greater funding, were assigned the best qualified teachers, and were able to recruit the highest scoring applicants on the national examinations; as a result, Tsinghua's stature grew steadily during the New Era.[25] Tsinghua's attached middle school also regained its stature as a keypoint school, and starting in 1978 it was able to recruit from among the top-scoring students on citywide middle school entrance examinations. During the transition period, students in the middle school were divided into "keypoint," "regular," and "basic" classes based on test scores, effectively separating the village and campus youth. Attention was focused on preparing those in the keypoint classes for the university examinations. The length of the keypoint program in the senior middle school was extended from two to three years and virtually all of the keypoint students tested into college, with many gaining admission to Tsinghua. Those in the regular and basic programs continued to graduate after two years and few tested into college. In 1981, the entire middle school was converted into a keypoint, and school officials opened a second school on campus to accommodate children of university employees who failed to test into the keypoint school. No provision was made for the graduates of primary schools in neighboring villages, who had attended Tsinghua's attached middle school during the Cultural Revolution decade.

While education officials concentrated on rebuilding the country's most elite schools, they also greatly reduced the size of the school system as a whole. Officials considered most of the rural middle schools opened in the 1970s, including those that had been created by expanding village primary schools, to be substandard. Between 1977 and 1983, over 105,000 rural middle schools were closed and the total number of middle school students dropped from

67,799,000 to 43,977,000.[26] As a result, the proportion of young people graduating from junior middle school dropped from over two-thirds at the end of the Cultural Revolution decade to just over one-third at the low point in the early 1980s; and the proportion graduating from senior middle school dropped from over 40 percent to less than 10 percent.[27] As the number of students was sharply reduced, the length of schooling was extended. Primary school was extended from five to six years, middle school from four to six years, and regular colleges from three to four years, while postgraduate programs were added to the top. Cultural Revolution policies had come close to creating a school system that was virtually flat, in which all children studied for nine or ten years; the post-Mao reforms reconstructed a much taller, but increasingly selective, education pyramid.[28]

Entrance examinations and keypoint schools once again became powerful mechanisms for reproducing the educated elite. Although the school system continued to provide a route to the top for a relatively small number of children from poor families who excelled in the examinations, the system grew progressively more closed. In sharp contrast with the past, when educators felt compelled to increase the proportion of university students who came from the countryside, university officials became increasingly concerned that students of rural origin lowered the quality of the student body. By the time I arrived at Tsinghua in the late 1990s, university leaders were unabashedly looking for ways to reduce the number of rural students. At that time, Chinese educators were in the midst of a campaign to improve the quality (*suzhi*) of education and of students.[29] This included efforts to reform the examination system so as to reduce its stultifying impact on middle school education and enhance the creativity of university applicants. There was wide agreement among education reformers at Tsinghua and other universities that one problem with the examination system was that it selected too many rural students, whose education had consisted of little more than intensive examination preparation. Many rural students who did well on the entrance exams, reformers argued, lacked the deeper intellectual qualities—manifested in foreign language competence, computer skills, musical training, and creative thinking—that students from urban keypoint schools brought to the university. Their explicit aim, therefore, was to reform the recruitment system so that it would select more students from urban keypoint schools and fewer from rural areas. In recent years, fewer than 20 percent of the students admitted to Tsinghua University have been from the countryside (where a large majority of the population resides), which is probably the lowest proportion since the beginning of the Great Leap Forward in 1958.[30] In the current ideological environment, however, even one in five students of peasant origin is considered too many.

REFORMING UNIVERSITY EDUCATION

Post-Mao education reformers moved quickly to restore the standards, credentials, curricula, teaching methods, and assessment procedures that had prevailed before 1966, eliminating all traces of Cultural Revolution policies that were designed to combine mental and manual labor. The new leadership at Tsinghua put an end to open door teaching practices, and students were no longer required to go to off-campus factories, farms, and construction sites as part of their studies. The number of campus factories was reduced from nineteen to nine, and production was cut back in the remaining factories, which were reoriented to serve more conventional research and teaching functions. Teachers and researchers returned from campus factories to their academic departments, and hundreds of workers were transferred out of the university. Some 1,200 workers who grew up in urban districts within the municipal borders of Beijing were sent to work in other local factories, while five hundred demobilized soldiers who had been hired by the university during the Cultural Revolution decade were sent back to their native villages.[31] According to Hong Chengqian, the campus factory leader, the ex-soldiers had no choice but to leave the university, but some of them also wanted to go. "Many of them felt that after the Cultural Revolution was over, their status in the university had fallen, so they were not happy being here."[32] Tsinghua's extension programs for industrial workers and its branch schools, including the Daxing County school for peasants, were closed or spun off to be administered by lesser educational entities.[33]

In line with central policies, the new Tsinghua administration exhorted university cadres and teachers to break with old "egalitarian" thinking that prevented the recognition and promotion of excellence. Progress, university leaders insisted, depended on identifying and cultivating students and teachers with outstanding talents. "Our thinking has still not been sufficiently liberated, our system has drawbacks, and egalitarian thinking is still a major obstacle," an article in the university newspaper lamented. "We must select the best, use the most talented, and emphasize our strong points, not our weak points."[34] The "teach according to ability" program was restored at the university and great emphasis was placed on new postgraduate programs. Tsinghua's attached middle school restored its special university preparation class, in which outstanding students received instruction from university professors and were accepted directly into Tsinghua University without having to take the examinations (in order to relieve them of tedious examination preparation and allow them to develop their creativity).[35] In 1988, the middle school's principal, Wan Bangru, accomplished his longtime dream, creating an experimental senior middle school class focusing on science composed of students recruited from around the country through special examinations.[36]

Jiang Nanxiang and other veteran officials who initially took the reins of the education system in 1977, were intent on restoring as perfectly as possible the status quo at Tsinghua and other universities before the Cultural Revolution, which had largely been modeled after Soviet practices. By the mid-1980s, however, university leaders moved to eliminate many features of the Soviet model of higher education in favor of American practices. Tsinghua, which had been restructured to focus on engineering in 1952, once again became a comprehensive university (although its strongest departments continued to be in the engineering field), and teaching was reorganized into broader disciplines, with general education and basic theory courses emphasized at the expense of practical training.

In the mid-1980s, in line with the central government's orientation toward market reform and global integration, Tsinghua opened a School of Law and a School of Economics and Management (SEM). The latter was founded by Zhu Rongji, the Tsinghua alumnus who later—as China's premier—would direct the privatization of much of the country's economy. The SEM, the first of its kind in China, became the largest and most popular school at the university. It was modeled after the leading business schools in the United States and it embraced the economic doctrines and business and management theories popular in those schools. Today, it offers MBA degrees and high-priced executive-training programs in collaboration with Harvard Business School and MIT's Sloan School of Management. In 2000, leading government officials, academics, and business executives from both China and abroad were invited to join a newly created advisory board. The foreign dignitaries on SEM's advisory board include top executives from Intel, Goldman Sachs, Nissan, GM, NASDAQ, McKinsey, BP, Temasek Holdings, Morgan Stanley, Merrill Lynch, Nokia, Citigroup, the Carlyle Group, and Blackstone.[37] Zhu Rongji serves as the honorary chairman of the board and the chairman is H. Lee Scott, Jr., president and CEO of Wal-Mart Stores, which in 2004 donated US$1,000,000 to SEM to help open the Tsinghua University China Retail Research Center.[38]

PLACING THE HEAD BACK ON TOP OF THE BODY

With the dawn of the New Era and the repudiation of the Cultural Revolution, the new CCP leadership renounced the party's long-standing effort to undermine the value of cultural capital and to undercut the social position of the old educated elite. As part of this effort, the party embarked on a campaign to repair its tattered relations with intellectuals. In the Central Committee's 1981 resolution on party history, which for the first time systematically criticized Mao Zedong, the new leadership harshly appraised the party's previous attitude toward intellectuals. "We must firmly eradicate such gross fallacies as the denigration of education, science and culture and discrimination against intellectuals, fallacies which had long existed and found extreme expression during the 'cultural revolution,'" the resolution declared. "We must unequivocally affirm that, together with the workers and peasants, the intellectuals are a force to rely on in the cause of socialism."[39]

In explaining the party's past hostility toward the educated classes, party spokespersons were apologetic, but they were also keen to stress that socialism would ultimately enhance the role of intellectuals. Hu Ping, a scientist who was given the task of articulating the CCP's new policies toward intellectuals, argued that in the past the party had been diverted from its true socialist mission. "Knowledge should be more required and intellectuals more respected in socialist society than at any other time in history," he wrote in a widely distributed 1981 essay. While feudal society had little use for specialized knowledge, and capitalists only valued intellectuals and their knowledge insofar as they contributed to profits, he explained, socialism would give full play to intellectuals and their specialized knowledge. Unfortunately, in the past the CCP had pursued an incorrect line toward intellectuals. This was not due to the nature of socialism, but rather because the party had emerged from base areas in the countryside, where peasants—who were engaged in small-scale production and had little use for science—were steeped in ignorance and superstition. "While mainly engaged in revolutionary wars, the cadres, unfamiliar with modern science, had no chance to raise their cultural level," Hu explained. As a result, for many years after 1949, intellectuals were persecuted, restricted, and cast aside, knowledge was despised, and "those without any expertise [were allowed to] lead those who are well-trained." Fortunately since 1978, Hu declared, "the party has made great efforts to correct past mistakes by reforming the political and economic management system and implementing a sound policy towards intellectuals."[40]

The new Tsinghua administration was very solicitous toward the senior professors who had been trained before 1949, and it sought to make amends for their previous mistreatment. As part of the effort to respect knowledge and talent and throw off previous prejudices, in the early 1980s the university rediscovered its venerable prerevolutionary heritage, and anniversary commemorations, books, and articles in a new alumni magazine paid tribute to the famous professors of the Republican era and celebrated the intellectual and social traditions of that period. In this nostalgic environment, the university's celebrated original entrance gate, which had been pulled down by Red Guards in 1966, was reconstructed. The new replica of the gate replaced a statue of Mao Zedong that the Jinggangshan faction had erected in its place in 1967.

China's educated classes were pleased that—after decades of disarray—the proper relationship between mental and manual labor was now being reestablished, both in terms of social status and economic remuneration. During the Mao era, when skilled workers were paid more than many college professors, intellectuals had complained that the "body had been placed on top of the head" (*naoti daogua*). Now the natural order had been restored, the head was back on top of the body, and educational achievement was more properly recognized in terms of social status and the distribution of economic rewards. From his new position as minister of higher education, Jiang Nanxiang insisted that past mistakes in the treatment of intellectuals had to be rectified. "We must carry out our intellectual policy in terms of first, political trust, second, economic security, and third, living and working conditions," he declared in 1979. "The outstanding problem today, is that intellectuals make less money than a worker of the same age."[41] This abnormality was corrected and in subsequent years the relationship was decisively reversed. Salaries at Tsinghua were adjusted with the aim of adequately compensating those at the higher rungs, especially middle-aged faculty, who had been denied raises since before the beginning of the Cultural Revolution. The professional-rank system was restored and almost three thousand teachers were promoted in rank and received corresponding increases in salary and benefits.[42] In addition to regular salaries, both Tsinghua University and its attached middle school were able to generate additional funds, especially after they began charging tuition, which were used to pay hefty bonuses to faculty members.

Soon after the workers' propaganda team was expelled from the university, Tsinghua cadres and professors were able to reclaim their original apartments, displacing the workers' families who had usurped their living quarters during the Cultural Revolution. In subsequent decades, the university built new apartments for many of its employees. At first, although the new apartments were constructed and distributed according to professional rank, there was little effort to create exclusive residential districts. In fact, in the 1980s when Tsinghua razed a nearby village to build university housing, it compensated villagers by providing them with apartment blocks located within the university's newly built housing complex. By the turn of the century, however, a new class hierarchy had been firmly established and such intermingling of social strata no longer seemed appropriate. In the late 1990s, when the university annexed Blue Flag Village just south of campus, it rejected villagers' demands that they be provided with local apartments, and those who refused to leave their homes were violently removed. The new complex of luxury apartments that replaced the village was reserved exclusively for top-ranking professors, researchers, and administrative cadres employed by Tsinghua and neighboring Peking University.

Chapter Ten

Triumph of the Red Engineers

On January 16, 1980, all of Tsinghua University's cadres at the department level and above were invited to the Great Hall of the People facing Tiananmen Square, where they joined thousands of other officials who had gathered to listen to a major speech by Deng Xiaoping. Tsinghua officials were excited about the changes that had transpired since Deng won political control, and his message on this day was cause for greater enthusiasm. "Expertise does not equal Redness," Deng declared, "but Reds must be experts" (*zhuan bu dengyu hong, danshi hong yiding yao zhuan*).[1] While these were encouraging words for members of the Tsinghua delegation, many others in the audience, Communist cadres who had little in the way of educational credentials, must have felt more than a little apprehensive. Deng's exhortation concisely expressed his goal of transforming the CCP, which had once been a party of peasant revolutionaries, into a party of experts. There had been tendencies in this direction in the past, but they had been impeded by the party's programmatic commitment to eliminating class distinctions. Mao had repeatedly mobilized party members and broader sectors of the population to resist technocratic tendencies, and had done so in a particularly determined fashion during the Cultural Revolution decade. Only after the Cultural Revolution had been repudiated and the party's class-leveling program had been abandoned, could Deng openly seek to make the CCP into a party of technocrats. In adopting this goal, Communist leaders moved decisively to cut the bonds that had long tied the party to the peasantry and working class, and to repair its ties to the educated elite.

With the removal of the class-leveling constraints of the Mao era, China's New Class established itself. This class represented an amalgamation of China's new and old elites, and it combined the political and cultural resources commanded by each. The groundwork for this convergence had already been laid during the previous three decades of contention and cooperation. First, the asset structures of the two groups had gradually converged, as children of peasant revolutionaries accumulated cultural capital, and children of the old educated elite accumulated political capital. Second, decades of contentious interaction had ultimately created the conditions for political unity between the old and new elites. The crucial impetus had been the Cultural Revolution, in which Mao unintentionally spurred inter-elite unity by simultaneously attacking both groups. After Mao's death in 1976, party officials and intellectuals found common cause in denouncing the violence and radical egalitarianism of the Cultural Revolution. Deng, who had led the campaign against outspoken intellectuals in 1957, now promised to make peace with the old educated elite. The CCP recognized the legitimacy of cultural capital without reservation, facilitating inter-elite convergence and the consolidation of a stable class order. Intellectuals became the prime target of party recruitment, and although the CCP continued to require political acquiescence, this was no longer so burdensome after the party abandoned class leveling and embraced a technocratic agenda that many intellectuals could readily accept as their own.

The heart of the New Class was made up of the Red and expert cadres who had been trained at Tsinghua and other universities during the Communist era. After 1976, the longstanding Red-over-expert power structure was dismantled; poorly educated revolutionary veterans were retired, and cadres who had both political credentials and advanced academic degrees—especially in engineering—were rapidly promoted to leadership positions. As a new class order based on political and cultural capital was consolidated, those who ended up at the top had both Red and expert qualifications, although the former had been drained of its original ideological meaning. This chapter will examine the process through which this new class hierarchy was consolidated.

Transforming the CCP into a Party of Experts

Under Deng, the CCP dramatically shifted recruitment to focus on intellectuals. Although the party had long tilted admissions toward the educated by making schools a key site of recruitment, it had also admitted large numbers of workers and peasants, and during the Cultural Revolution decade recruitment efforts had been deliberately shifted down the social hierarchy. As a result, at the close of the Mao era, the educational qualifications of the great majority of party members were very low. According to statistics gathered in 1985, the bulk of party members were still from the poorly educated classes: 10 percent of party members were illiterate, 42 percent had only a primary school education, and 30 percent had graduated from junior middle school; 14 percent had graduated from senior middle school, and only 4 percent had graduated from college. By that time, however, party leaders had already moved to fundamentally change the class composition of the party. The proportion of new members categorized as intellectuals grew from 8 percent in 1979 to about 50 percent in 1985.[2]

In 1978, the system of class designations was abolished. This was a tremendous relief for old elite families that had endured landlord, rich peasant, capitalist, and other "bad" class labels and had faced severe discrimination in political recruitment, school admissions, and job placement. It was also a relief for intellectuals who did not suffer the taint of bad class origins, but who had been hampered by the party's class preferences favoring workers and peasants and its general distrust of the educated classes. Now all intellectuals were explicitly given preference in recruitment and promotion to positions of leadership. By the same token, peasants and workers no longer received official favor and the old class preferences were now tainted by association with the Cultural Revolution; those who had been promoted during that period were subsequently viewed with disdain and suspicion. Although the party continued to recruit workers and peasants, the clear implication of Deng's maxim that all Reds should be experts was that there was no longer much room in the party for members of the laboring classes, and there was little prospect they would become leaders.

The new Tsinghua leadership enthusiastically implemented directives to recruit intellectuals. During the Cultural Revolution decade, the workers' propaganda team had focused on recruiting campus workers and had regarded faculty members with suspicion. The university party organization now reversed course, inviting veteran professors who had long been denied membership to join the fold. Table 10.1 presents the distribution of party membership among Tsinghua University employees by occupation and rank in 1993. The figures reveal an employment hierarchy in which educational and political

TABLE 10.1
Party membership among Tsinghua University employees, 1993

Employee status and rank		Total	Party members	Party members' proportion of total
Faculty	Professors	701	540	77.0
	Associate professors	1,323	894	67.6
	Lecturers	786	406	51.7
	Assistant teachers	531	168	31.6
	Subtotal	3,341	2,008	60.1
Administrative staff		484	261	53.9
Specialized staff	High	399	193	48.4
	Middle	843	313	37.1
	Low	705	93	13.2
	Subtotal	1,947	599	30.8
Workers		1,805	326	18.1
Total		7,577	3,194	42.2

SOURCE: Fang and Zhang (2001, vol. 1, 819).

credentials largely coincided: at each rung, both rates of party membership and educational credentials increased in tandem. The difference between the rates of party membership among the teaching faculty (60 percent) and among workers (18 percent) was dramatic. Just as illuminating was the correlation of professional rank and party membership *within* the faculty and specialized staff. This relationship had been reversed since the 1950s, when very few professors were members. Moreover, at that time there were no senior faculty members among the core leadership of the university party organization, which was composed of veteran revolutionaries, including peasants and leaders of the underground student movement. By the early 1990s, the university party organization had become a "professors' party" and its leadership was made up of senior faculty.

The same kind of pattern emerged in all work units, and the change was most dramatic in factories. In the 1960s, as noted in Chapter 3, factory party organizations were overwhelmingly made up of workers, who were far more likely than technical cadres to join the party. Now the party was much more interested in recruiting technical cadres than workers, and the main base of factory party organizations was increasingly located in administrative and technical offices rather than on the shop floor.[3] In a national survey of urban residents, Andrew Walder found that during the Mao era there was little difference in the rates of party recruitment between those who had a college degree and those who did not, but after 1987 college graduates were nearly six times more likely to join the party.[4]

College graduates came to expect that those with prestigious educational credentials should also have commensurate political credentials, and vice versa. Han Lingzhi recalled that as a Tsinghua graduate she was expected to be a party member. "We were from Tsinghua and when we got a job we would be the backbone employees (*gugan*), so we had to be party members," Han—who joined the party before she graduated in 1982—told me. "Let's say you go to some kind of office or research institute; the other people from other universities, they are party members already and I'm from Tsinghua and I'm not. Definitely I'm much better than them, so it's just embarrassing. If you're academically better, usually every aspect is better."[5] Those who had succeeded in both the academic and political competitions were understandably proud of their accomplishments. As university graduates and party members, they were part of two highly select groups, each of which carefully distinguished itself from the general population. Imbued with the thinking of the New Era, they were confident they merited the positions of authority and privilege to which their combined credentials now entitled them.

Red Experts Take Charge

During the 1980s, the Red-over-expert power structure that the CCP had maintained since taking power in 1949 was dismantled. As recounted in Chapter 1, this structure had been established when the CCP sent its own cadres, typically soldiers of peasant origin, to take charge of enterprise and government offices that were staffed by White experts inherited from the old regime. For decades, the structure had been reproduced as the CCP established separate technical and administrative career tracks and generally placed university graduates, whom it continued to regard with suspicion, in technical positions with little power. In factories, the party preferred to recruit managers and party cadres from the ranks of the workers, and the most capable were promoted to positions of leadership in the enterprise, and then in municipal, provincial, and ministerial bureaucracies. During the Mao era, university-trained engineers—even if they were party members—were generally not considered leadership material. The role of party leaders, in contemporary thinking, was to mobilize the workers and peasants, a task best suited to cadres who emerged from the ranks of the masses and spoke their language. The role of engineers, in contrast, was to provide technical assistance. In the New Era, mass mobilization was replaced by "scientific management," and engineers were now considered to have the best qualifications to be administrative and political leaders.[6] University graduates, who had previously been relegated to technical positions, were now promoted to positions of power: engineers became factory directors and party secretaries; researchers became directors of ministerial bureaus; and planners became mayors and municipal party secretaries.

During the Mao era, especially during its most radical moments, the CCP also had preferred to recruit technical cadres from among the workers. During the Cultural Revolution decade, Tsinghua had been deeply involved in this task in three ways: by training factory workers who would return to their work units after three years of education; by helping large factories establish their own July 21 universities; and by conducting short-term technical classes for workers from smaller factories. Under Deng, the CCP closed down the July 21 universities and put an end to the practice of filling technical positions from among the ranks of a factory's own workers. Now all new cadres—technical and managerial—were to be recruited directly from schools. In 1980, Hu Yaobang, the party's general secretary, made the new policy explicit, declaring, "[We should] recruit cadres from the graduates of colleges, middle schools, or equivalent ones. [We should] generally not directly select [cadres] from among workers and peasants who have little education."[7] As a result, class position was decided in the education system: those who did not get very far in school became workers and remained workers; those who tested into higher schools become cadres.

Even the military, which had long been an alternative route—outside the school system—for accumulating political credentials, was now changing. Previously, the People's Liberation Army had promoted officers from its own ranks, selecting promising recruits—almost all of whom were peasant youth—

for political cultivation and officer training. Since many military officers later transferred to the civilian realm, this also had been an important path by which peasant youth, who faced daunting odds in the academic credentialing system, became leadership cadres in civilian institutions. Starting in the early 1980s, candidates for officer training schools were selected from among senior middle school graduates who did well on admissions examinations. The officer corps underwent a technocratic transformation, as the People's Liberation Army enhanced its elite educational apparatus, creating a National Defense University and developing officer training programs at Tsinghua and other top universities.[8]

Not only was mobility from noncadre (worker, soldier, etc.) positions to cadre positions now much more restricted, but the demarcation between the two was more clearly defined. Previously, many workers who had been promoted to technical and managerial positions (including workers who had served as factory and department leaders at Tsinghua) had retained their formal status as workers and their original pay grade (which was often higher than that of their new cadre position). This ambiguous status, referred to as "worker-acting-as-a-cadre" (*yigong daigan*), was now eliminated; these workers either returned to their former production duties, or they were formally given cadre status and received cadre pay and benefits, which now rose much faster than those of workers. In the New Era, there was also little call for cadres to participate in productive labor or for workers to participate in management; these were increasingly seen as outmoded practices that hindered efficiency and scientific management.

During the 1980s, much of the existing cadre corps was replaced in what Cheng Li and Lynn White called "probably the most massive tranquil elite transformation in history."[9] The new regime systematically replaced poorly educated cadres, both revolutionary veterans and worker-peasant cadres promoted during the Mao era, with cadres who were, in Deng's words, "younger, better educated, and better qualified professionally."[10] As Hong Yung Lee noted, by replacing poorly educated veteran cadres with better-educated younger cadres, the government not only improved technical competence, but also put in place a cadre corps more sympathetic with Deng's technocratic policies. Many older party officials resisted the changes in recruitment and promotion policies. Press reports cited by Lee criticized cadres who were "not yet freed from the ossified thinking of the leftists," and who, therefore, regarded young candidates as "only expert, but not Red," "arrogant," "detached from the masses," "seeking the bourgeois lifestyle," and "immature and unstable." These veteran cadres continued to see intellectuals as "targets of reform, [who] can be used but not trusted," and they claimed that "if they are recruited, the party will change its characteristics."[11] The momentum of reform, however, would not be stopped by such complaints; instead, the criticisms were used as evidence that these cadres were part of the problem the campaign was designed to eliminate.

Between 1982 and 1988, 1,630,000 cadres who had joined the Communist movement before 1949 retired; during the same period, another 3,120,000 cadres who had been recruited after 1949 also retired.[12] Many of these cadres were reluctant to retire and only left as the result of an unrelenting party campaign, the explicit aim of which was to get rid of cadres who did not meet Deng's requirements. In 1986, the *People's Daily* reported with satisfaction that during the previous six years more than 469,000 college-educated cadres had been promoted to leadership positions above the county level.[13] In just three years, from 1982 to 1984, the proportion of municipal leaders with college degrees increased from 14 percent to 44 percent, and the proportion of college-educated county cadres increased from 14 percent to 47 percent.[14] At the top, the proportion of college graduates on the party's Central Committee increased from 26 percent in 1977, to 55 percent in 1982, 73 percent in 1987, 84 percent in 1992, 92 percent in 1997, and 99 percent in 2002.[15]

A New Technocratic Class Order

During the 1980s, the CCP constructed a social system and a class hierarchy very much in accord with the elitist Saint-Simonian vision of socialism. Although the Marxist vision of eliminating class distinctions had been abandoned, this order—like that in the contemporary Soviet Union—remained socialist in the sense that it was still based largely on public property. Private enterprise was severely restricted until the early 1990s, and the class order that was established during the intervening years is of particular theoretical interest because it was based largely on political and cultural capital, and the role of economic capital was still very limited.

During this period, the class hierarchy was based first and foremost on the advantageous position of the public sector compared to the private sector. In cities, the public sector—state and collective enterprises, government offices, schools, hospitals, and so forth—was dominant, and private enterprise was only permitted at the margins, mainly in small-scale retailing and services. In rural areas, land remained village property, but was farmed by individual families, and there was a rapidly growing sector of township and village enterprises, which harbored both socialist and small-scale capitalist production relations. Because private entrepreneurs, urban and rural, were not allowed to legally hire more than seven employees, family labor prevailed in the private sector.

In the public sector, class position was synonymous with rank in the administrative hierarchy, and career advancement was determined largely by possession of political and cultural capital. In fact, the more the public sector embraced bureaucratic principles, the more it made a fetish of both academic and political credentials. In the family-labor sector, on the other hand, academic and political credentials were far less common, but they were also less important. Very few peasants or self-employed urban residents were party members and few had senior middle school diplomas, much less college degrees. But the success or failure of household economic activities did not depend on either. Vocational skills and social and political connections were important, as were location and luck, but even more important were entrepreneurial abilities, and the road to success was marked not by the accumulation of credentials and appointment to positions in a bureaucratic hierarchy, but rather by accumulation of property. Because of restrictions on the private sector, the paths to the top of the class hierarchy were located within the public sector and those who had the educational or political credentials needed to do well in the public sector generally preferred the formal career tracks it offered. In fact, the private sector was composed largely of people who were excluded from the public sector.

At the top of the public sector, which in 1990 employed more than 230 million people throughout China, were some twenty-eight million cadres.[16] Educational credentials were now the defining asset that distinguished cadres from the masses of workers below, as access was determined by how well an individual did in school; graduates from colleges and higher-level technical schools continued to be guaranteed cadre jobs. These positions were still divided into two tracks, one technical and the other political/administrative, and the party organization at every level remained in charge of appointments to key administrative posts. Although only a minority of all cadres belonged to the party, the top 345,000 or so "leadership" positions—those ranked at the county or division chief level and above—were reserved almost exclusively for party members, and by that time promotion into the ranks of this upper stratum typically required a university degree.[17] These were the Red experts who ran the country.

At the dawn of the Communist era in the 1950s, two distinct elites—very different and mutually suspicious—had faced each other in the upper levels of Chinese society. Four decades later, the top echelons of society were occupied by a much more homogenous stratum of Red experts. Many of these Red experts were offspring of new elite families, while others were offspring of old elite families, but it was no longer so easy to tell them apart. Some had stronger political credentials and others had stronger educational credentials, but they all had both. The next generation was even less distinct. Grandchildren of illiterate peasant revolutionaries and highly cultured patricians attended the same elite schools, and although it was still possible to distinguish traces of distinct inherited advantages and disadvantages, these were slight compared to earlier decades. Some students could cite an intellectual pedigree that went back generations, while others could recount their family's revolutionary legacy, but both had educated parents and both were free to pursue political careers without discrimination. In place of contending elites, a New Class had come into existence. This class had its roots in both of the erstwhile elites, and its ranks had been replenished—through the political and academic credentialing systems—both from within and from below.

Red Engineers Ascend to the Top

At the Sixteenth CCP Congress in 2002, the reins of the party and state apparatus were passed to a set of leaders, with Hu Jintao at the head. In the party's reckoning, this was the fourth generation of leaders since the CCP had come to power in 1949. Previous generations had all joined the party while it was still an insurgent organization; this was the first generation that had come of age in the Communist era and had been selected and trained by the political and academic credentialing systems scrutinized in this book. All nine members of the Political Bureau's Standing Committee, the most powerful men in the country, had been trained as engineers, and four, including Hu, were Tsinghua alumni.

The reasons for the extraordinary preeminence of engineers, and Tsinghua graduates in particular, among the CCP's top leadership can be found in the party's history. The party had placed great emphasis on training engineers ever since it reorganized China's universities and government ministries in accord with the Soviet-inspired rapid industrialization project in the 1950s. Not only had the majority of university graduates specialized in engineering, but engineering majors had attracted students with the highest examination scores. Then, after Deng came to power, all leadership cadres—from ministers and provincial party secretaries down to county officials—were expected to be at the forefront of scientific administration, planning, and modernization, and engineering training was associated with the outlook and aptitude most appropriate for leadership.

Because under Jiang Nanxiang's leadership in the 1950s and 1960s Tsinghua University had established its reputation as China's top engineering school and built a political recruitment organization that was held in high regard, after 1978 party leaders turned to Tsinghua alumni from the Jiang era to find young cadres with leadership potential. Tsinghua graduates who had been active in the university party organization, serving as student cadres and political counselors, had earned prized academic and political credentials that now promised rapid upward mobility. Early in the post-Mao era, the university cemented its position as the country's premier training ground for leadership cadres, and as a result the number of Tsinghua graduates among China's most powerful officials was far greater than the number of graduates of any other university. More than three hundred of the university's graduates have served as ministers or deputy ministers of the State Council, and thousands of others have served as factory directors, bureau chiefs, mayors, governors, and local and provincial party secretaries.[18] Because so many Tsinghua alumni have climbed to the top rungs of the political establishment, they have been collectively referred to as the "Tsinghua clique" (*Qinghua bang*). Despite the terminology, Tsinghua graduates have never formed any sort of political faction, but many have tended to favor hiring and promoting graduates from their alma mater and they are linked by strong social networks. As a result, as more alumni have reached positions of power, the value of a Tsinghua diploma has steadily increased.

By reviewing the biographies of the key CCP leaders over the last three decades it is possible to trace the party's technocratic transformation and follow the ascent of Red engineers—from Tsinghua as well as from other leading technical schools—to the top of China's political hierarchy. The diverse personal backgrounds and histories of these individuals also provide a glimpse of the tumultuous process by which old and new elites converged. The leading members of the second generation that took the reins in the late 1970s, including Hu Yaobang and Zhao Ziyang, were revolutionary veterans who had joined the movement in the 1930s. Protégés of Deng Xiaoping, they were patrons of the party's technocratic transformation, but they personally had relatively little education or technical training. The subsequent generations have all been dominated by Red engineers.

The leading members of the third generation, which came to power in the late 1980s, all joined the underground Communist movement during the 1940s, but they also received training as engineers. Jiang Zemin, who served as the party's general secretary, and Li Peng, who served as premier and head of China's National People's Congress, were the sons of revolutionary martyrs and were raised in Communist households. They were both assigned by the party to study engineering and after finishing their training in the Soviet Union in the early 1950s, they embarked on careers that started in factories and led to important positions in the machine-building and electric power ministries.[19] On the other hand, Zhu Rongji, who was named deputy premier after the Fourteenth Congress and later succeeded Li as premier, was from a very wealthy Hunan landowning family that reportedly could trace its genealogy to Zhu Yuanzhang, founder of the Ming dynasty. Zhu tested into Tsinghua under the old regime, where he became an activist in the underground student movement and joined the CCP in 1949. After graduating with an electrical engineering degree in 1951, he worked for the State Planning Commission until 1958, when he was dismissed and thrown out of the party after being accused of having Rightist tendencies.[20]

All three men rose extraordinarily rapidly after 1978. Jiang was asked to head the State Commission on Foreign Investment in 1980, briefly served as chief of the Ministry of the Electronics Industry, and was then named Shanghai's party secretary, before being elected as the party's general secretary in 1987. Li was asked to head the Ministry of the Power Industry in 1979, became deputy premier in 1983, and was then appointed acting premier in 1987. After Zhu's party membership was restored in 1978, he returned to planning work and by 1983 he was deputy minister of the State Economic Commission. Zhu succeeded Jiang as Shanghai's party secretary in 1987, returned to Beijing to become deputy premier in 1991, and then replaced Li as premier in 1998. While head of state, Zhu continued to serve as dean of Tsinghua's School of Economics and Management, which he had founded in 1984.

The fourth generation, which rules China today, was born in the 1940s and attended university and joined the CCP after 1949. The careers of the three men who today occupy the top positions in China's party and state apparatus followed very similar trajectories. In college they all excelled academically and were exemplary student cadres, but during the first years of their careers—during the Cultural Revolution decade—they were largely confined to work in the technical realm. Hu Jintao, who became the party's general secretary in 2002 and was later named president and chairman of the Central Military Commission, was born into a clan of wealthy tea merchants from Jiangsu Province. After testing into Tsinghua's water conservancy engineering department in 1959, he became a top-ranked student and an enthusiastic political activist. He led the schoolwide dance troupe, joined the party, served as a political counselor, and was selected to be a political instructor. As a cadre in Jiang Nanxiang's political recruitment apparatus, he became an early target during the Cultural Revolution. Former students reported that Hu, like most other student cadres, was sympathetic with the moderate camp during the ensuing factional conflict, but he wisely refrained from becoming deeply involved. After the workers' propaganda team arrived in 1968, he was sent to work on a construction team in western Gansu Province, and following a year of manual labor, was assigned to technical positions designing and building hydroelectric projects.

Wen Jiabao, China's premier, came from a family of schoolteachers in Tianjin. Choosing to follow his father's specialization in geology, in 1960 he tested into the Beijing Institute of Geology, where due to his political activism he was one of a select group of students invited to join the party. After graduating in 1968 he was also sent to Gansu Province, where after doing a stint of manual labor he spent most of the next decade working as a field geologist. Wu Bangguo, the chairman of the National People's Congress, is the son of a basic-level cadre who taught in the People's Liberation Army Institute of Surveying and Mapping. Wu was accepted into Tsinghua's radio electronics department in 1960, and, like Hu, he joined the party and worked as a political counselor. After graduating, he was assigned to work in an electronic tube factory in Shanghai, where after a year of manual labor he became a technician.

Although Hu, Wen, and Wu were largely relegated to technical work during the Cultural Revolution decade, early in the reform era—after it was decided that engineers should be in charge—they were identified as candidates for rapid promotion. At first moving up in the economic bureaucracies where they began their careers, they rapidly advanced into positions of political power. Hu climbed the ranks in Gansu's Ministry of Construction, but then in 1980 he was transferred to the Youth League, and four years later was named national secretary, joining the top echelons of the central party leadership in Beijing. He then served stints as secretary of the provincial party committees in Guizhou and Tibet, before returning to Beijing in 1992 to head the Central Party School, an appointment that signaled that he was being groomed for the party's top position. Wen first rose in the ranks of Gansu's Geology Bureau; then in 1982 he transferred to the Ministry of Geology in Beijing and was named deputy minister the following year. In 1985, he moved to the party's General Office and became its director a year later, establishing his position among the top leadership of the party. Wu was appointed director of his electronic tube factory in 1978, and he subsequently rose quickly in Shanghai's industrial bureaucracy, joining the Standing Committee of the Municipal Party Committee in 1983, and then succeeding Zhu Rongji as party secretary in 1991. Later that decade, both Wen and Wu were appointed to deputy premier positions.[21]

By the time Hu, Wen, and Wu took over the top party and state posts after the Sixteenth CCP Congress in 2002, the party leadership was completely dominated by Red engineers. The proportion of the party's ruling Political Bureau that was made up of individuals with science and engineering degrees had grown dramatically, increasing from none in 1982, to 50 percent in 1987, 75 percent in 1998, and 76 percent in 2002. By then, as noted above, all nine members of the Political Bureau's Standing Committee had been trained as engineers.[22] The proportion of engineers in the party's top leadership bodies has since declined, and it seems that the fifth generation will include more individuals with training in other fields. Indeed, of the nine new members appointed to the Political Bureau at the CCP's Seventeenth Congress in 2007, only three have engineering degrees, and the others have degrees in economics, management, planning, law, math, history, and philosophy. Of these new members, it seems that Xi Jinping is being groomed to take Hu Jintao's place as general secretary in 2012, and Li Keqiang is being groomed to take Wen Jiabao's place as premier. Xi, the son of a prominent party official, attended Tsinghua's chemical engineering department as a worker-peasant-soldier student in the 1970s and later received a doctorate in political education from the university. Li, the son of a basic-level cadre, tested into Peking University after the examinations were restored in 1977, earning an undergraduate law degree and eventually a doctorate in economics. Significantly, neither man has any experience working in a technical field. Because they graduated from college in the first years of the New Era, when the Red-over-expert system was being dismantled, they were—unlike their predecessors—able to directly pursue political careers. Xi moved into government administration and Li became a national leader of the Youth League.[23]

Should the fact that engineering degrees and technical work experience are not as common among the fifth generation of party leaders be interpreted as a sign that the party's technocratic character is eroding?[24] Such an argument might be made, but only in a narrow sense: as central planning has given way to market exchange and China has become more integrated into global political and economic systems, the educational credentials regarded as qualifications for leadership have evolved, and the value of training in economics, management, law, and other fields has increased relative to the value of training in science and engineering. The essential elements that make the party technocratic, however, remain in place. Leaders continue to be systematically selected and prepared by the academic and political credentialing systems that have long underpinned China's technocratic order, and the newly built schools of economics, management, and law at Tsinghua and other universities have proven as adept as science and engineering programs at training young technocrats. Moreover, the methods of selection and training have become even more elaborate and refined, as postgraduate education and party school training are required for career advancement, and the entire apparatus continues to rigorously select for competence, bureaucratic efficacy, and technocratic values, grooming officials who share the party's prescribed "scientific outlook on development."[25]

Chapter Eleven

Technocracy and Capitalism

On June 27, 1997, enterprising professors and cadres at Tsinghua University eagerly awaited the results of the initial public offering (IPO) on the Shanghai Stock Exchange of Tsinghua Tongfang Company, Ltd. They had just created the company, combining elements of the stable of small research facilities and factories the university had maintained since the 1950s. The IPO was a success and the value of the company's stock tripled on the first day. Since then, due in part to acquisitions of other enterprises, Tsinghua Tongfang has grown spectacularly: between 1997 and 2006 its annual revenues increased from 47 million to over 12 billion yuan (US$1.5 billion). In addition to developing one of China's leading brands of personal computers, the company provides information technology services and manufactures a growing range of other products, including flat-screen televisions, air-conditioning and lighting systems, inspection technology, wastewater treatment equipment, optical disks, pharmaceuticals, and military communications systems.[1]

Tsinghua Tongfang is a hybrid public-private corporation. The university retains 34 percent equity and the remaining shares are owned by other investors, including the company's key executives, who might be called state sector entrepreneurs. The chairman of Tongfang's board of directors is Rong Yonglin, who also heads Tsinghua Holdings Company, the entity created to manage the university's multibillion-dollar investment portfolio. Although Rong received a chemical engineering degree from Tsinghua, his subsequent career at the university has been political and administrative. He joined the CCP a year after he graduated in 1970, and later served in various capacities, including manager of the university's computer factory, secretary of the schoolwide Youth League committee, and assistant to the president. In contrast, Tongfang's chief executive officer (CEO), Lu Zhicheng, followed the "professor-turned-CEO" trajectory that the university now encourages. Lu was admitted to Tsinghua as a worker-peasant-soldier student in 1974, and earned a civil engineering degree in 1977 and a master's degree in thermal engineering in 1983. As a Tsinghua professor, he did research on computer-controlled air conditioning and commercially developed the technology by starting a university-owned enterprise, which later became part of Tongfang.[2] Hu Haiqing, the thirty-six-year-old who heads Nuctech, a Tongfang subsidiary that specializes in X-ray inspection technology, is one of the company's rising stars. Hu is both well connected and well trained: his father is President Hu Jintao and he received a master's degree in engineering physics from his father's alma mater.[3]

It would be hard to exaggerate how much Tsinghua's enterprises have changed over the last three decades. In the 1970s, when Rong Yonglin was first appointed to be assistant manager of the university's computer factory, which at that time was producing China's first desktop computers, his salary probably did not exceed 80 yuan a month. That was less than what many of the skilled workers in the factory made, and he was expected to "live, work, and eat" with the workers. Today Tongfang's top executives receive annual salaries of up to 450,000 yuan, plus bonuses and stock dividends, and they probably never talk to and seldom see the workers who assemble Tongfang computers in the company's numerous production facilities. During a 2000 interview, Lu Zhicheng, Tongfang's CEO, proudly introduced a *China Daily* reporter to Sun Jiaguang, a Tsinghua professor who had converted a software innovation into a Tongfang stock option, becoming an "instant millionaire." Lu predicted that within three to five years, Tongfang would produce "one thousand millionaires."[4]

Tongfang's headquarters are located in a black glass-and-steel edifice that towers over the academic buildings on the Tsinghua campus. The building is part of Tsinghua Science Park, a sixty-two-acre complex that might be described as the university's new capitalist face. The park, which opened in 2000, provides offices for twenty multinational corporations, including NEC, Sun Microsystems, Toyota, Microsoft, Google, P&G, and Schlumberger Technologies, as well as scores of Chinese high-tech companies, including many of the enterprises in which Tsinghua Holdings, the company that manages the university's investment portfolio, has a stake. It is also the site of Tsinghua's Returned Students Pioneer Park. Pioneer Park was created to "incubate" start-up high-tech companies; it provides office space, legal and financial consulting, seminars on business management, and networking opportunities to budding entrepreneurs, most of whom are Tsinghua alumni who have returned to China after studying and working abroad. Tsinghua Holdings provides venture capital to those deemed most promising.[5] Ventures similar to Tsinghua's high-tech companies and incubator programs can be found at universities across the country, and they are part of a much larger phenomenon: China's Red engineers, or at least some of them, are becoming capitalists.[6]

Starting in the early 1990s, sweeping market reforms—including privatization and the elimination of lifetime employment—restructured the Chinese economy along capitalist lines. Deng Xiaoping's highly publicized tour of foreign-funded enterprises in southeast China's Special Economic Zones in early 1992 is conventionally cited as the key moment that marked the shift to more radical economic reforms.[7] This chapter will describe the subsequent capitalist transformation of China's economy and consider how the reintroduction of economic capital has altered a social hierarchy that had been based largely on political and cultural capital.

Capitalist Transformation

After 1992, the CCP strongly encouraged the growth of the private capitalist sector and by the end of the decade it had also privatized the great majority of publicly owned enterprises. Most state-owned and collective enterprises became the property of their managers.[8] Between 1991 and 2005, the proportion of the urban workforce employed in the public sector fell from about 82 percent to about 27 percent.[9] The state held on to the largest and most important enterprises, particularly those in the banking, oil, steel, power, telecommunications, and armaments industries, but their structure was fundamentally changed so that they were required—and able—to make profit generation their primary goal. To accomplish this, they shed their previous obligations to their employees, removing themselves from the business of providing housing, health care, child care, recreation, education, and other services for employees and their families.[10] Lifetime-employment guarantees were eliminated, and it has been estimated that by 2002 over fifty million workers—about 40 percent of the public enterprise workforce—had lost their jobs due to restructuring.[11] Enterprises reduced the size of their workforce, but they also discharged veteran workers and replaced them with younger workers who were less costly and more pliant. State-owned coal mines, for instance, now engage contractors who compete to mine coal for the lowest cost per ton, using migrant labor.[12] As a result of these reforms, the socialist sector of the Chinese economy has ceased to exist; virtually all enterprises—including those that are owned in part by the state—now operate according to capitalist principles.

The introduction of large-scale private property is reorganizing the Chinese class structure. The class of Red experts, which consolidated its position at the top of Chinese society in the 1980s, is being transformed, and the dominant class that eventually emerges from this transformation will not be composed of the same individuals and will not have the same bases of power. Many Red and expert cadres, such as the engineers who run Tsinghua Tongfang, have converted themselves into successful entrepreneurs. Indeed, a large proportion of China's fledgling capitalists have emerged from inside the party-state establishment. This is true, on the one hand, because the great bulk of the nonagricultural economy is comprised of companies that began as government entities, from industrial giants to humble township and village enterprises. On the other hand, many of the private enterprises that suddenly mushroomed in the 1990s were also created by individuals who had connections inside the establishment. These included professionals and managers who left public sector careers, as well as the relatives of public sector cadres. In fact, children of party officials make up a disproportionately large contingent of the new entrepreneurial class, and the pattern of officials' children going into business extends from village party branches up to the top echelons of the party.[13]

At the very top, political power has certainly spawned economic success. The children of the three men most responsible for engineering the privatization of China's state-owned enterprises in the 1990s, Jiang Zemin, Li Peng, and Zhu Rongji, have all done very well in the business world. Jiang's son, Jiang Mianheng, who has a Ph.D. in electrical engineering from Drexel University in Philadelphia, is deputy president of the Chinese Academy of Sciences and political commissar of the People's Liberation Army's General Equipment Department, where he has focused on information technology transformation in the military. At the same time, he has, together with Winston Wang, the scion of one of Taiwan's wealthiest industrialists, founded Grace Semiconductor Manufacturing. The company has invested hundreds of millions of dollars in building semiconductor plants in Shanghai (and it has retained Neil Bush, the U.S. president's brother, as a consultant in exchange for US$400,000 annual compensation, paid in company stock).[14] Li Peng's son, Li Xiaopeng, and his daughter, Li Xiaolin, who are both Tsinghua graduates, each heads an enormous state-owned power company, which together produce as much as 15 percent of China's electricity.[15] The children of Zhu Rongji have both gone into banking; his daughter, Zhu Yanlai, is manager of the planning department at the Bank of China in Hong Kong, while his son, Zhu Yunlai, is CEO of China International Capital Corporation, China's largest investment bank.[16] The children of China's current top leaders, Hu Jintao and Wen Jiabao, are all involved in the information technology industry in one capacity or another. Wen's son, Wen Yunsong, is CEO of Beijing Unihub Global Network, a major IT services provider, and his daughter is general manager of Great Wall Computer Corporation, one of China's largest computer manufacturers.[17] Hu's son, as we have seen, heads a division of Tsinghua Tongfang, and in 2003 his daughter, Hu Haiqing, who—like her father and brother—earned an engineering degree at Tsinghua, married Mao Daolin, one of China's wealthiest Internet magnates. As CEO of Sina.com, a popular Internet portal, Mao had accumulated shares worth at least US$67 million.[18]

It would be wrong, however, to think that the emerging capitalist class is simply the metamorphosed descendant of the Red expert elite. Only a small part of China's Red and expert cadres have become entrepreneurs, and many of those who have taken the plunge have achieved only modest success. Of the thousands of cadres and professors at Tsinghua, for instance, only a few have become high-tech entrepreneurs, or capitalists of any type, and the same is true across the public sector. The set of qualities that make a successful Red and expert official are not altogether the same as those that make a successful capitalist. Indeed, many of the most successful of the new generation of Chinese capitalists started out at lower rungs of the social hierarchy. They include headstrong managers of humble cooperative factories who today run industrial behemoths and ambitious peasant entrepreneurs who have amassed huge fortunes in construction and real estate development.[19] Only after the privatization process has finished turning public into private wealth, and all of the outstanding niches in the market have been filled, will we be able to assess the extent to which the capitalist class that emerges is the genetic offspring of the Red and expert officials who set the privatization process in motion.

From the top to the bottom of the social hierarchy, businesspeople are generally wealthier than state officials, but they have less lofty academic and political credentials. If we compare the individuals at the very top of the public and private sectors, leading public sector officials virtually all belong to the party and have university degrees, while only a minority of those on the lists of China's wealthiest men and women (most of whom are associated with entirely private enterprises) are party members or college graduates.[20] These distinctions are a product of the structural differences between the public sector, where political and cultural capital still rule, and the private sector, where entrepreneurship and private property play a more important role. Nevertheless, we can expect the differences between elite families that emerged from the public and private sectors to diminish over time, as families associated with the public sector scramble to convert their political and cultural capital into economic capital, while families that have succeeded in business despite relatively weak political connections and a lack of advanced degrees endeavor to use their economic resources to obtain both.

The Impact of the Return of Economic Capital

Beyond the problem of elite circulation and reproduction, the rise of capitalist enterprise raises questions about institutional changes in the bases of class power. How has the transformation of China's economy changed the technocratic order that emerged in the 1980s? How has the reintroduction of economic capital affected the operation and relative importance of political and cultural capital?

CHINESE CAPITALISM AND THE RESILIENCE OF POLITICAL CAPITAL

Since the CCP came to power in 1949, the value of political capital has been based on the party's control of China's state apparatus and the economic organizations it commands. The party has now surrendered much of the economic field to the private realm and has committed itself to allowing the remaining state enterprises to operate according to capitalist economic principles. As a result, in comparison with the days of the planned economy, the party's power has declined substantially, and with it the relative importance of political capital. Nevertheless, political capital remains a central mechanism of class differentiation. It functions in distinct ways in the public and private sectors. In the public sector, including Tsinghua University, the CCP continues to control the personnel system and it has formal jurisdiction over appointments to key administrative posts. A successful administrative career in the public sector, therefore, continues to require party membership and selection for party school training. In addition to these impersonal political credentials, the party organization continues to serve as a framework for cultivating personal networks that are critical for success.

Like the imperial bureaucracy of the past, the CCP rotates state officials to prevent the development of local coalitions that might undermine central power; for this purpose, the party shuffles top executives among airlines, telecommunications companies, and other key state enterprises, and shifts top officials from government posts to enterprise management and vice versa. Western economists object that the party's influence over personnel decisions in state sector enterprises violates international norms of corporate governance by introducing political considerations into decisions that, in principle, should be based on shareholders' interests. CCP control over appointments will undoubtedly face growing opposition from international actors as well as state sector managers and other domestic investors who have financial stakes in particular enterprises, but so far the party has shown little inclination to relinquish this key lever of power.[21]

In the private sector, political capital is also important, but it functions differently. In this sector, the party organization is sparser, personnel offices are not controlled by party committees, and there is no dossier system or *nomenklatura*. As a result, party membership has much less value as a bureaucratic credential. Instead, its value lies in its function as a networking tool. At every level, the party organization serves as a link to other members of the political, economic, and social elites. For those in business, it provides connections not only to government officials, but also to other successful entrepreneurs. For this reason, many private entrepreneurs find it advantageous to be party members, and the party is eager to recruit successful entrepreneurs; the larger the enterprise, the more likely the owner is to be a member.[22] Moreover, surveys have indicated that a high proportion of managers in private enterprises are party members and that when private entrepreneurs interview job applicants they consider party membership a plus.[23] Thus, even students at Tsinghua's School of Economics and Management who intend to look for a lucrative position in the private sector after graduation have reason to join the party. As long as the CCP is able to retain control of the state apparatus, political capital derived from association with the party will be worth a great deal, both as a credential and a networking tool.

Today, Red engineers continue to govern China, using a technocratic state apparatus to regulate a dynamic and unruly capitalist economy. Ultimately, it is the party's control over the appointment of officials in the public sector that continues to give the Chinese state its technocratic character. The party is fundamentally committed to technocratic principles and it runs a personnel system that—through rigorous competitions—selects and trains officials with technocratic credentials and values. The party has preserved the state's technocratic character by keeping two forces at bay. First, it has stifled threats from below that might bring to power a populist agenda, and second, it has been able to contain the political influence of private capital's narrow pecuniary interests. Both popular demands and economic corruption continue to threaten the centralized power and organizational integrity on which the technocratic character of the state depends, but so far the party organization has held.

THE APPRECIATION OF CULTURAL CAPITAL AND THE FADING OF THE MERITOCRATIC VISION

On the one hand, the impact of capitalist transformation on cultural capital can be gauged in terms of the monetary value of academic credentials. Just considering this very concrete measure, the value of cultural capital has increased dramatically as a result of the market reforms of the 1990s. Before then, during the early years of the reform era, although educational achievement was rewarded more handsomely than it had been in the Mao era, the salary scale in the public sector was still relatively compressed and cadres' salaries were fairly modest. For instance, in the late 1980s the base salary for full professors at Tsinghua University was 160 yuan a month, compared to about 40 yuan for newly hired cafeteria workers, both rates set by national regulations. The actual income gap, including bonuses and unequal provision of housing and other fringe benefits, was wider but still relatively narrow. Because of the caps on public sector salaries, many professors were upset that outside the university gates there were uneducated peasant entrepreneurs who were making more money than they did (although few had any desire to abandon public sector careers and try their luck marketing fresh produce). The reforms of the

1990s assuaged the educated classes by creating a more open labor market in which the value of academic credentials soared. The emergence of a large-scale private sector, which included the newly opened offices of multinational corporations, offered a lucrative employment alternative, and salary caps in the public sector were lifted. A national study of urban employees' salaries showed that the value of having a college degree (compared to a senior middle school diploma) in terms of added annual income more than tripled between 1988 and 2001.[24] At Tsinghua, the salary gap today is huge. With inflation and the commodification of housing and other goods and services, it is difficult to directly compare salaries today with those of twenty years ago, but the differences are astonishing: many workers in Tsinghua's cafeterias today make the minimum wage of 580 yuan a month, while the university brags that it pays top-notch professors over one million yuan a year. In Gary Becker's terminology, freeing the labor market has substantially increased the return on investment in human capital.

The growing income gap at Tsinghua reflects nationwide trends. In 1978, China's Gini coefficient, a measure used to compare international income inequality in which 0 indicates absolute equality and 1 indicates absolute inequality, was calculated to be 0.22. This was among the lowest rates in the world, and scholars were particularly impressed because of China's size and geographic diversity. Indeed, China had accomplished the low rate despite huge income differences between urban and rural areas and between developed and undeveloped regions because within each locality differences were minimal. Less than three decades later, in 2006, China's Gini coefficient was calculated to be 0.496, surpassing the United States and approaching the rates of the world's most unequal countries, such as Brazil and South Africa.[25] Inequality between regions and inequality between rural and urban areas have both increased substantially, but the most dramatic change has been the polarization of income within localities, and—as we have seen at Tsinghua—within workplaces. Much of the increase in inequality since 1992 can be attributed to income derived from property ownership, but the appreciation of cultural capital has also played a very important role. The market value of academic credentials has benefited from the growth of large-scale private enterprises; even public sector professionals, including professors and cadres at Tsinghua, can now demand higher salaries and perks, justifying their claims by pointing to escalating standards in the private sector. Over the last decade, as their salaries have grown, the best compensated of Tsinghua's professors have begun abandoning their relatively modest university apartments to join private entrepreneurs in suburban gated communities and luxury high-rises that have mushroomed in Beijing.

While the pecuniary value of cultural capital has grown substantially, however, the radical market reforms of the 1990s had a deleterious effect on grand meritocratic visions that flourished in the 1980s. The decade between 1978 and 1989 was the heyday in China of meritocratic theories, according to which admission into the top ranks of society should be determined primarily by knowledge. As engineers and scientists were being promoted to positions of leadership throughout the vast state apparatus, meritocratic ideas captured the imagination of Chinese intellectuals. Astrophysicist Fang Lizhi and rocket designer Qian Xuesen, nationally renowned scientists who were appointed to positions of prominence, eloquently articulated widely held views about the leading role of intellectuals. "Since we say that intellectuals are the leading force, responsibility for China thus falls on our shoulders," declared Fang, challenging his audiences to claim the place that history had bestowed upon them.[26] Qian proposed that by the year 2000, all cadres should have a university degree, all cadres at the county or bureau level and above should have a master's degree, and all ministers and governors should have a doctoral degree.[27] Some saw the CCP as the vehicle for this kind of technocratic transformation; they insisted that modernization required not only technical expertise, but also the firm guidance of a centralized and efficient political authority. Others saw the party as an obstacle because it continued to consider political loyalty in promoting individuals, shortchanging knowledge and ability, and they urged the party to give up its monopoly on power. Opinions on this subject stimulated lively debates among university students, faculty members, professionals, scientists, engineers, managers, and officials inside and outside of the party.[28]

These debates in China in the 1980s were reminiscent of those that had gripped Eastern Europe starting in the 1960s. In Konrad and Szelenyi's introspective account, old guard party officials were stubbornly trying to hold on to their special pretensions to power based on political qualifications, while the broader intelligentsia was pushing for a more meritocratic system. Eastern European intellectuals, they wrote, were pursuing a "New Class project," the ultimate aim of which was to make cultural capital the main determinant of class power. This goal might be described as a *pure* technocracy, unsullied by political intrusion. In the 1960s and 1970s, visions of a technocratic and meritocratic future were popular not only in the socialist East, but also in the capitalist West. While Konrad and Szelenyi and others were forecasting the retreat of political power and the triumph of cultural power in the East, prominent Western scholars, including Daniel Bell, John Kenneth Galbraith, and Alvin Gouldner, were concluding that the reign of capital was giving way to social stratification based on knowledge. The world's future seemed to belong to an ascendant "knowledge class."[29]

In China, hopes that political power would retreat were shattered by the harsh repression of the student movement in 1989. After the Tiananmen crackdown, the CCP continued its own technocratic transformation, promoting experts to positions of authority and making the state more technocratic than ever, but it has shown no inclination to surrender power. Moreover, in recent years as wealthy entrepreneurs claim their places at the top, meritocratic visions—in which social selection would be based primarily on education and intellectual ability—have gradually faded. With the country increasingly in the thrall of money, the idea that society might soon be ruled by a knowledge class seems more and more far-fetched. Even dissident intellectuals who are impatiently awaiting the demise of the CCP no longer entertain such pretensions. In the end, meritocratic prophesies succumbed to a foe more formidable than tanks—economic capital.

The technocratic class order that emerged in China in the 1980s, when public property still prevailed, was underpinned by the marriage of political and cultural power; now political power has found a new partner. The profound changes that have taken place in China in recent years are, of course, part of a global phenomenon. In the 1990s, economic capital reasserted its primacy with a vengeance, not only in China but around the world. For the time being at least, public property and economic planning have been sidelined, undermining the foundations of the grand meritocratic visions of the past. As the state has retreated from the economy, allowing markets and profit margins to increasingly dictate the allocation of resources, the dream of creating a Saint-Simonian society, with an economy governed rationally by scientists on behalf of the public, has quietly wilted. In its place, a technocratic state does its best to regulate an economy that increasingly belongs to Mammon.

TSINGHUA UNIVERSITY IN THE NEW CLASS ORDER

Every year at the end of April, tens of thousands of Tsinghua alumni return to their alma mater for a gala celebration commemorating the founding of the university in 1911. The formal festivities on campus are just a backdrop for banquets and meetings that bring together former classmates, friends, and associates who are scattered across the country and around the world. Some are engineers in small enterprises or county offices, while others command huge bureaucracies as municipal party secretaries, ministerial bureau chiefs, or corporate executives. The most prominent participants include top government officials (Zhu Rongji and Hu Jintao have frequently topped guest lists), chief executives of China's largest state-owned enterprises, and the fabulously wealthy owners of some of the country's most successful high-tech start-ups. Virtually all of those who gather in April are members of the Tsinghua Alumni Association, which has chapters in every significant Chinese city, as well as in cities across the globe. A small but growing number of alumni are also members of the Tsinghua Entrepreneur and Executive Club (TEEC), which grew out of these annual alumni gatherings and holds a conference at the university every April. Many of the club's founders made their initial fortunes in California's Silicon Valley, and today the organization's members include millionaires and billionaires from both sides of the Pacific. Members donate large sums of money to the university and return to the school to participate in an annual Future Entrepreneur Training Camp. "We do this out of deep feelings towards Tsinghua," declared Yang Lei, TEEC deputy president and a high-tech entrepreneur. "It's kind of a reward to [the] mother school by helping our younger schoolmates."[30]

Tsinghua's alumni associations are formal manifestations of the manifold social networks that connect graduates of the university. These Tsinghua networks, together with other university alumni networks, help tie together the new dominant class in China, from top to bottom. The most influential members of the Tsinghua alumni networks derive their power from their positions in the CCP. They began to establish critical connections while they were student members of the Tsinghua party organization, and the personal networks they have cultivated around the governing party constitute what Bourdieu would have called embodied political capital. Others derive their power mainly from the economic capital at their disposal and the business networks they have cultivated, networks that are also greatly enriched by Tsinghua connections. The fact that these political and economic networks revolve around an academic institution is one indication of the density of the links that tie together political, economic, and cultural capital in today's China.

Conclusion

The twentieth-century Communist endeavor to eliminate class distinctions was the most ambitious and traumatic social-leveling experiment in human history. Leveling movements have punctuated all of recorded history, but none has surpassed the Communist movement in terms of duration, geographic scope, the number of people involved, and the level of ideological sophistication. Although it is difficult to precisely date the end of this experiment, the events of 1989 and the subsequent restoration of large-scale private property in most of the countries of the Soviet bloc and China was certainly a decisive moment. Although a number of Communist parties remain in power, almost all have brought back private property, abandoning an essential element of the Communist program and opening the way for a rapid polarization of wealth and income. By this reckoning, then, the 1990s marked the demise of the twentieth-century Communist project. Communist efforts to eliminate class distinctions, however, had come to an end much earlier. Long before the return of private property, Communist parties in the Soviet Union, China, and elsewhere presided over well-established class hierarchies based on public property, with a new class of Red experts on top. These countries were still in some sense socialist, but they had abandoned Marx's vision of a classless society in favor of Saint-Simon's vision of a society governed by the talented.

Although these Communist experiments were ultimately a spectacular failure, there is much that can be learned from this failure. There will inevitably be more social-leveling movements in the future, and although private property is held sacrosanct by constitutions around the world today, it is hard to imagine that those without property will long tolerate a small minority of the population owning the great bulk of the wealth. Future social-leveling movements will undoubtedly revive demands for public ownership, and that is reason to look more closely into the origins and nature of the class hierarchies that the Communist experiments with public ownership produced. In this book, I have focused on class advantages derived from possession of cultural and political assets, and have examined conflicts surrounding the institutions that distributed these assets. I have shown that, despite promises to do away with class differences, the Chinese Communist Party, like its counterparts in other socialist countries, became a determined champion of class differentiation based on political and cultural capital. In this concluding chapter, I will return to the questions posed at the beginning of the book: when Communist revolutionaries dispossessed the propertied classes and created socialist systems based on public property, was their intention to build a technocratic class order? If not, was this result inevitable?

To lay the groundwork for answering those questions, I will first compare the Chinese with the Soviet experience, identifying goals, conflicts, processes, and results that seem to be common to the Communist project and others that seem to be peculiar to China. The aim is twofold. First, because the Soviet Union established the model that the CCP and other Communist parties followed, understanding the origins of this model will shed light on what happened in China. Second, a comparison of the results of the two most important Communist experiments will better enable us to answer the general questions posed above. In the following pages, I will first review the early history of the Soviet Union, extracting from a number of insightful studies those elements that have particular bearing on the subsequent Chinese experience, and I will then summarize the findings of the present inquiry.

The Circuitous Soviet Road to Technocracy

In 1917, the Bolshevik Party and the Russian technical intelligentsia had important potential points of unity. As Kendall Bailes noted, both were determined to use science to modernize Russia, and the Bolsheviks' dirigisme was not entirely foreign to scientists, engineers, and planners who had been employed by the czarist state. Moreover, Lenin admired the knowledge and practical orientation of the technical intelligentsia, and in the chaos that followed the fall of the Russian monarchy—with armed workers taking charge of factories and peasants seizing land—many of the technical intelligentsia preferred the Bolsheviks, with their penchant for order and discipline, over some of their populist and anarchist competitors.[1] If the Bolshevik movement had been inspired by a Saint-Simonian vision, the two groups might have collaborated to create a technocratic society in relatively short order. The Bolsheviks, however, were instead committed to the Marxist goal of eliminating class distinctions. This goal put them sharply at odds with members of the educated classes, including the technical intelligentsia, and set up a protracted period of sharp conflict.

Relations between the Communist Party of the Soviet Union (CPSU) and the educated elite in the early years of the Soviet era were far more volatile and antagonistic than those between the CCP and the educated elite in China in the early 1950s. The first six months were particularly difficult, as a good part of the educated classes refused to cooperate with the fledgling revolutionary regime. The Bolsheviks responded to widespread strikes by decreeing compulsory labor service for all technical specialists; Lenin called the confrontation a civil war between the Soviet authorities, on one side, and representatives of the bourgeoisie and sections of the intelligentsia, on the other.[2] During this period, the Bolsheviks created a Red-over-expert administrative structure like the one later instituted in China. The party sent political commissars to supervise incumbent managers and technical personnel, and although it paid the technical intelligentsia well for their cooperation, it shunned them politically, instead reinforcing its own ranks by recruiting workers and promoting them to positions of authority. In 1928, ten years after the Bolsheviks seized power, there were only 138 engineers in the party, compared to 742,000 workers.[3] Soviet factories were typically run by a Red director, who usually was a former worker and seldom had more than a grade-school education; this Red director relied on the assistance of a general engineer, who usually was not a party member.[4]

In Russia, as in China later, early Communist education policies fluctuated but moved generally in a radical direction. Lenin made Marx's comment about the need to eliminate the distinction between mental and manual labor into a fundamental goal of the new socialist state, and the education plank of the Bolshevik program, drawn up in 1919, promised to transform the school "from the weapon of bourgeois class domination into a weapon for the total destruction of class divisions within society."[5] The program promised to provide free compulsory education for all children up to age seventeen, throw open the doors of universities to all, and establish a close connection between study and productive labor.[6] These, of course, were promises for the future; at the time, most of the population was illiterate, there were few schools, and most teachers were hostile to the new regime. Schools, which remained largely autonomous for the first decade of the Soviet era, became the site of competing agendas. Soviet education authorities strove to implement progressive teaching methods based on practical learning and all-around intellectual development, many teachers tried to maintain conventional standards and methods, and proletarian students fought to take control of schools away from "bourgeois" teachers and students. In postsecondary schools, a wide social and political gap separated students from old elite families, who had graduated from traditional high schools, from working-class students, most of whom had been recommended by trade unions, the party, and the Communist Youth League to attend special workers' schools to prepare for college.[7]

In 1928, the CPSU—now under Stalin's leadership—moved decisively to the Left, embarking on agricultural collectivization and rapid industrialization, and commencing a period of cultural radicalism that Sheila Fitzpatrick has called the Soviet Union's "Cultural Revolution."[8] Rapid expansion of primary and secondary education was accompanied by a massive adult literacy campaign. Postsecondary education also expanded rapidly, with emphasis on short-term technical programs and narrowly defined majors that quickly prepared graduates to join the industrialization drive. Grades, exams, and conventional lectures were criticized, and political classes, practical learning, and student participation in productive labor were stressed. Class discrimination against students from old elite families, which had begun with the Bolshevik takeover, was intensified, and 65 percent of admissions were reserved for workers and peasants and their children. Entrance examinations were eliminated and some 150,000 adult workers, a large proportion of whom were party members, were recommended to attend higher-level schools. This wave of proletarian students was urged to finally take control of schools out of the hands of bourgeois professors, by monitoring teachers, challenging their authority, and participating in school administration.[9]

Radical education policies coincided with attacks on educated elites in all sectors, including industry and government. Factory workers were encouraged to challenge the authority of specialists, and the contributions of worker-inventors and worker-technicians, who had only on-the-job training, were celebrated as superior to those of engineers. After the Shakhty affair in 1928, in which a group of mining engineers were tried for "wrecking," workers were encouraged to closely monitor the experts in their factories, and several thousand were arrested. In another highly publicized trial in 1930, a group of top-level engineers and experts in Soviet ministries and planning offices were accused of forming an "Industrial Party" and conspiring to replace the Bolshevik government with one run by technocrats.

TAKING THE TECHNOCRATIC ROAD

In 1931, Stalin delivered two speeches that marked the end of this period of cultural radicalism and the beginning of what Nicholas Timasheff famously called the "Great Retreat."[10] In the first, Stalin denounced "petty-bourgeois egalitarianism," and in the second he announced that the old intelligentsia was coming over to the side of the Soviet government.[11] Radical education policies were gradually reversed. Recommendation of adult workers for higher education was curtailed, entrance examinations were restored, and school fees were introduced. Grades, course exams, and conventional methods of teaching were restored, and practical learning and student participation in labor declined. Students were admonished not to interfere in administration and encouraged to respect the authority of teachers. Class preferences were eliminated and attacks on the old educated elites, whether professors or engineers, were discouraged.[12] Professional ranks were restored and salaries and benefits were increased for those who had higher qualifications and made greater contributions, leading to a steadily increasing gap between workers and technical and managerial employees.[13]

In the early 1930s, hundreds of thousands of newly minted proletarian Red experts graduated and entered the labor force. They had originally been recruited to go to school so they could take the place of unreliable bourgeois experts, but in the end most of the old experts were retained. It turned out, after radical policies were discarded, that there was room enough for both old and new experts in the ongoing industrialization drive. Nevertheless, many of the young Red experts were promoted to take the place of the poorly educated Red directors and managers (who had been recruited in the 1920s from the ranks of the workers), a process facilitated by the massive purge of officials in 1936–38.[14] As a result, the Red-over-expert structure was replaced by a model in which political, administrative, and technical leadership were combined and authority was concentrated in the hands of a single Red and expert director. Members of the first cohorts of Red experts rose quickly in the party and state bureaucracies during the Second World War and the postwar years, and they came to dominate the Soviet leadership in the 1960s and 1970s.[15] Because of the strong class preferences of the late 1920s and early 1930s, these leaders were typically of working-class origin, but after class discrimination was eliminated, students were drawn increasingly from among children of the educated classes.[16]

Although political movements would continue to claim hundreds of thousands of victims in the years to come, after cultural radicalism was renounced in the early 1930s, major political campaigns did not specifically target members of the old educated elite or make social inequality connected with education a cause for political mobilization.[17] In 1936, Stalin declared that Soviet society was composed of two friendly classes, the working class and the peasantry, and a stratum he called the "working intelligentsia."[18] This latter term, as Fitzpatrick pointed out, not only replaced the pejorative "bourgeois intelligentsia," but now encompassed both members of the old intelligentsia and party cadres.[19] By the time Soviet advisors arrived in China in the 1950s, the Soviet Union was administered by Red engineers, who had been selected by a decidedly hierarchical school system and a party organization dedicated to technocratic principles.

The goals of achieving a classless society and eliminating the distinction between mental and manual labor remained an official part of CPSU doctrine—and were consecrated in the country's constitution—until the Soviet Union's demise. After cultural radicalism was renounced in the early 1930s, however, this goal was to be accomplished not by means of class struggle, but rather through the gradual cultural improvement of the workers and peasants so that they reached the level of the intelligentsia, who now served as a model for emulation. This new interpretation provoked much less social turmoil, but it did little to diminish class differences. During subsequent decades, the salaries and material conditions provided for members of the intelligentsia grew steadily compared to those of workers, and the institutions that underpinned the social advantages of the educated elite and facilitated its reproduction across generations were elaborated and fortified.[20]

China and the Soviet Road

In many ways, conditions in post-1949 China were much more propitious for technocratic development than those in the early years of the Soviet Union. Relations between the new Communist regime and the educated elite were much less confrontational as a result of several factors, including the CCP's more secure hold on power, its governing experience in rural base areas (where it had substantial practice cooperating with local elites), and its comparatively moderate approach toward urban elites. This approach was encouraged by the new regime's Soviet advisors, who came with blueprints for building socialism that reflected the technocratic turn that had taken place in the Soviet Union years earlier. Nevertheless, the CCP rejected the technocratic road, and instead embarked on a class-leveling project that closely resembled that undertaken in the early years of the Soviet Union.

The Chinese ending up re-creating many of the same policies that had been implemented in the Soviet Union decades earlier, but had since disappeared (even in official histories). No systematic scholarship has yet been done on the extent to which radical Chinese policies during the Great Leap Forward and the Cultural Revolution were modeled after early Soviet policies.[21] To whatever extent they were, it is clear that the fundamental impetus for both was the same: Marxist ideas as adapted by Lenin. The CCP was a product of the extraordinary global appeal of Leninist ideas following the October Revolution. These were ideas that appealed to revolutionaries, not technocrats. Leninist doctrine insisted on violent revolution and unremitting class struggle; its adherents were instructed to organize the most impoverished and oppressed of the workers and peasants to fight for a classless society. The CCP waged nearly three decades of rural warfare under Leninist banners, and it brought these ideas, along with a huge army of peasant cadres, to power in 1949.

The Chinese Communists, like the Bolsheviks, converted the means of production into public property, and control was concentrated in state and collective offices, access to which was provided by political and cultural capital. The latter was concentrated in the hands of the old educated classes, while the former was concentrated in the hands of party cadres. If class leveling had stopped with the redistribution of economic capital and left the existing distribution of cultural and political capital undisturbed, the stage would have been set for the relatively tranquil development of a technocratic order. Instead, class leveling was extended to the cultural field, and eventually to the political field. In pursuing class leveling, the Chinese followed the Leninist theory and practice they had learned in the 1920s and 1930s, and to a great extent they were prisoners of this theory and practice, even when they extended their use into uncharted territory.

REDISTRIBUTION IN THE CULTURAL FIELD

The program that the CCP inherited from the Bolsheviks, as we have seen, called for eliminating not only private property in the means of production, but also the distinction between mental and manual labor. The concentration of education in the hands of the old privileged classes was seen as no more morally justified than the concentration of property. The willingness of Communist cadres to carry out radical redistribution policies in the cultural field was reinforced by the fact that the overwhelming majority—who were poorly educated peasant revolutionaries—had few personal assets at stake. They took the same kind of class-struggle approach in the cultural field that they took in the economic field, mobilizing the disadvantaged against the advantaged. Although cultural capital could not be confiscated and redistributed the way economic capital was, they redistributed educational opportunities, restructured the institutions that reproduced the unequal distribution of capital, and undermined the social authority and status of the educated elite. Although the Communists rapidly expanded the school system, they were not content with gradually improving the conditions of the lower classes; instead they were determined to tear down the advantages of the educated classes.

In both the industrial and education fields, the CCP initially adopted the technocratic models pioneered by the Soviet Union after the CPSU abandoned class leveling. In industry, it attempted to implement the Soviet "one-man management" model, which combined political, administrative, and technical leadership. In education, it created a meritocratic hierarchy of schools, which advantaged children of the educated classes. These policies had support within the party and, as we have seen, technocratic ideas had a great deal of appeal at Tsinghua and other universities that were charged with training Red engineers. Nevertheless, technocratic policies ran into tremendous resistance and were ultimately rejected, as the CCP moved to pursue cultural class leveling in an increasingly radical fashion.

The one-man management model, which placed expert managers in charge, was never popular with most Chinese Communist cadres, who had little technical expertise themselves and did not trust the incumbent experts. Instead, they preferred to keep power in the hands of party committees, and the CCP formally rejected one-man management in 1956.[22] When the party took over factories (as well as government offices, schools, and other institutions), it established a Red-over-expert power structure, as the fledgling Bolshevik regime had done, in which incumbent managers and specialists were relegated to subordinate technical positions. Over time, this Red-over-expert structure was reinforced and reproduced because the CCP distrusted new university graduates (who were largely from old elite families) and preferred to promote workers to positions of power.

Starting in 1957, the CCP also rejected technocratic aspects of the Soviet education model and implemented much more radical policies. These policies were partially reversed following the collapse of the Great Leap Forward, but Mao revived the radical agenda in 1964, and cultural-leveling initiatives reached their height during the Cultural Revolution decade. The radical education program in China was very similar to radical policies that had been implemented in the early years of the Soviet Union, but the CCP carried it out in a more systematic fashion and for a longer period of time. The main goals were: (1) to redistribute educational opportunities, inhibiting the reproduction of the educated elite and dispersing education across the population; and (2) to alter the nature of education and of occupational divisions so as to eliminate the distinction between mental and manual labor. Hundreds of thousands of schools were built, and the length of primary and secondary education was reduced as part of a crash campaign to make both universal. All middle school graduates entered the workforce, starting in manual occupations, and postsecondary education, which could only be provided for a minority, was reorganized to hinder the reproduction of the educated elite. Entrance examinations were eliminated, students were recommended by factories, communes, and military units, and class line discrimination against old elite families was intensified. Short-term training programs—located increasingly in factories and villages—were developed and curricula were designed to combine theory with practical knowledge and manual skills, preparing graduates for occupations that would combine mental and manual labor.

By the end of the Mao era, class distinctions based on education remained enormous, but they had been significantly diminished. Radical education policies had many deleterious side effects, but they were effective in advancing their class-leveling goals. This conclusion contradicts Jonathan Kelly and

Herbert Klein's oft-cited thesis that revolutions cannot redistribute cultural assets. Based on studies of Bolivia and Poland, they concluded that radical social revolutions were capable of initially reducing inequality by redistributing physical property, but failed in the long run because they were incapable of reducing inequality based on education.[23] In the Chinese case, although the CCP certainly did not eliminate inequality based on cultural capital, during the Mao era such inequality never ceased to give ground to radical assaults. Ultimately, the fundamental obstacles that stymied class leveling in China were in the political, not the cultural field.

REDISTRIBUTION IN THE POLITICAL FIELD

While the CCP considered the cultural field—and the education system in particular—to be enemy territory, the political field was Communist territory. The party sought to capture this enemy territory and redistribute enemy resources, and at the same time shore up its own position in the political field by concentrating power in the hands of the party. Communist cadres saw themselves as leading the workers and peasants in a struggle against the old elite classes. Under this banner, the CCP reorganized villages, factories, and schools, making them into highly organized work units with semipermanent members, and within each unit it endeavored to concentrate power in the hands of a party committee. Thus, in the political field it came to power with an agenda of concentrating, not dispersing power. Nevertheless, the party was concerned that its own cadres were abusing the power they had and becoming estranged from the masses. To carry out a revolutionary agenda directed against the most powerful groups in society, it depended on support among the lower classes, and it required cadres who acted as revolutionaries-in-power, not privilege-seeking officials. It therefore attempted to employ the masses to monitor its own cadres. Party work teams were sent to villages, factories, and schools to mobilize work unit members to criticize local leaders for "bureaucratic" behavior—isolation from the masses, abuse of power, "commandism," and suppression of criticism from below. Communist cadres were subjected to periodic campaigns—all initiated by Mao—against corruption and bureaucracy, including the Party Rectification campaign in 1957 and the Socialist Education movement in the early 1960s. Like campaigns in the cultural field, these campaigns became larger, longer, and more disruptive over time, culminating in the Cultural Revolution.

In the political field the Chinese also started with Soviet theory and practice. They borrowed from the Soviets not only the Leninist party model, with its vanguard status, hierarchical structure, and strict discipline, but also methods of rectifying problems in the party organization. The CCP's concept of bureaucracy—and the struggle against it—came from the Soviets. Chinese Communists learned to practice criticism and self-criticism and organize mass supervision of cadres from their Soviet mentors, and Chinese leaders were following Soviet precedent when they issued strongly worded denunciations of party officials for abusing workers and peasants, and when they discovered enemies within the party bent on restoring capitalism.[24] Nevertheless, both the goals and methods of the Chinese Cultural Revolution extended the Bolshevik struggle against bureaucracy in ways that made it qualitatively different.

Mao identified the target of the movement as an emergent bureaucratic class that was exploiting the workers and peasants. He was convinced that party officials had become the main danger to the Communist class-leveling project, and he adopted goals and methods that reflected this concern. While previous campaigns had targeted official abuse of power, the Cultural Revolution was the first that was clearly designed to disperse cadres' political power. The main goals were: (1) to redistribute political power within work units, by undermining the authority of cadres and enhancing the power of rank-and-file work unit members; (2) to weaken patterns of political tutelage and patronage; and (3) to prevent cadres from obtaining privileged access for their children to party membership, education, and employment. Mao went around the party apparatus and used his personal authority to spur the creation of an autonomous rebel movement. This movement, with Mao's support, effectively undermined the authority of the party bureaucracy, enhancing power at the top and the bottom at the expense of party officials in the middle. Mao gained much greater personal authority, and the masses—or at least some of them—enjoyed unprecedented power as local rebels mobilized people to criticize local party officials and decide who among them was fit to return to office.

After fighting between rebel groups and moderate defenders of the status quo brought China to the brink of civil war, Mao authorized suppression of the factional conflict. The rebel organizations were disbanded and the party—which had been paralyzed for two years—was rebuilt, reconcentrating power in the hands of its officials. Mao, however, attempted to check bureaucratic power by creating a system of institutionalized factional contention that pitted rebels against administrators. Veteran cadres were returned to administrative positions, while rebels were placed in positions that allowed them to mobilize opposition to the administrators. Radical leaders continued to mobilize political movements against party officials, but—unlike the rebels of the early years of the Cultural Revolution—they employed administrative measures and bureaucratic methods of mobilization. At Tsinghua, the workers' propaganda team mobilized students and workers to criticize university cadres and teachers. Although the system of governance they created prevented the restoration of an orderly bureaucratic hierarchy, it ended up perpetuating a culture of political tutelage, this time in a distorted form I have called sycophantic rebellion.

During the Cultural Revolution, Mao continued to insist on putting politics in command, that is, enhancing political at the expense of cultural power by giving precedence to political over technical and academic considerations and qualifications. His strategy was in essence to transfer power from the cultural to the political field, but at the same time to disperse cadres' political power. Ultimately, the fate of this strategy—and of the entire class-leveling project—rested on finding effective means to disperse cadres' political power. Although the early rebel movement was very effective in undermining the authority of party offices, the results of subsequent efforts to institutionalize factional contention and mass supervision over cadres were disappointing. Ultimately, the political experiments of this period provide little indication that Mao and his followers had found effective means of dispersing the power of Communist officials.

REPRODUCTION AND CONVERGENCE OF ELITES

So far, I have discussed the cultural and political fields separately, but China's New Class arose at the intersection of the two. This class was the product of the reproduction and convergence of the political and educated elites. The political elite was able to reproduce itself as their children acquired political credentials, and the educated elite was able to reproduce itself as their children acquired educational credentials. It was the convergence of the two groups, however, that created a technocratic class.

Structurally, the political and educated elites converged gradually as their asset structures became more similar. In 1949, there was little overlap; very few members of the educated classes belonged to the party, and very few party members had higher levels of education. Those who had a foot in both camps—the Communist intellectuals—were tiny in absolute numbers and a small minority within each group. As children of the educated elite gained political credentials and children of the political elite gained academic credentials, the number of people who occupied the intersection of the two groups—the Red experts—grew steadily. Their twin credentials gave them a vested interest in the hierarchical structure of both the cultural and political fields, and their common experience in higher education and Communist activism imbued them with common values and perspectives that set them apart from the great majority of the population, including most of their parents.

The two elites also converged politically as members of both groups came to recognize a mutual interest in preserving social stability and halting class-leveling campaigns. During the first decades of Communist power, these campaigns were facilitated by the gulf between the new and old elites. Communist cadres saw the educated elite as representatives of the old order and believed that undermining the privileges they derived from cultural capital was part of the party's revolutionary mandate, while intellectuals saw Communist cadres as unqualified usurpers and resented the privileges they derived from political capital. In 1957, members of the two groups lined up on opposite sides of battle lines defined by political and cultural capital. In 1966, the same kind of inter-elite antagonisms exploded at many elite schools, but simultaneous attacks on both groups ended up forging inter-elite unity. One manifestation of this unity was the moderate factions that emerged at Tsinghua University and other schools. Budding Red experts of all class origins took umbrage at radical slogans denouncing party-affiliated college graduates as "new bourgeois intellectuals" and they came together in the moderate camp to defend both political and cultural capital.

This unity could not be consummated during the Cultural Revolution decade, when Mao and his radical followers throttled all elite pretensions. After Mao died in 1976, however, party officials and intellectuals discovered unprecedented unanimity in condemning the violence and egalitarianism of the Cultural Revolution. By then, the gradual convergence of old and new elites had established the conditions for the rapid consolidation of a technocratic class. Party leaders renounced class leveling, unambiguously recognized the value of cultural capital, embraced the old educated elite, and moved to transform the CCP into a party of technocrats. Intellectuals now acclaimed Deng Xiaoping—who had organized the persecution of dissident intellectuals during the Anti-Rightist campaign two decades earlier—as their savior. The party organization and the education system were refurbished and Red experts were moved expeditiously into positions of responsibility, replacing veteran peasant revolutionaries and worker-peasant cadres. The new CCP leadership, like their Soviet counterparts, insisted they were not abandoning egalitarian Communist ideals, only the destructive approach of class leveling. In the future, everyone would get rich, but some would get rich first. As it turned out, after the institutions that underpinned class differentiation were rebuilt, they were reinforced, and social inequality increased steadily.

Communism, Technocracy, and New Class Theory

Because the Communist revolutions in Russia, China, and other countries ended up creating a technocratic order with Red experts on top, it is tempting to think that this was the plan all along. Konrad and Szelenyi presented this idea in a particularly compelling fashion, developing a narrative in which the Bolsheviks and other Communist parties were the vanguard of the intelligentsia and the architects of a technocratic order that fulfilled intellectuals' long-held ambitions for power. In other words, according to Konrad and Szelenyi, Communist revolutionaries from the beginning sought to realize a Saint-Simonian vision of socialism. In this book, I have considered the Chinese case and tell a much different story. The CCP finally did take the technocratic road, but only after it abandoned the road of class leveling. This change of course converted the party from the enemy into the champion of cultural capital, and it facilitated the consolidation of a New Class based on the convergence of old and new elites. It seems to me that the basic elements of this story—the abandoning of class leveling in favor of technocratic policies and the contentious convergence of old educated and new political elites—also fit the Soviet case, and they are likely to fit other cases in which Communist parties came to power by means of indigenous revolutions. Any general theory that seeks to explain the rise of a New Class in socialist societies, I believe, has to make a place for these elements. With these elements in place, New Class theory would then have to explain the changing orientation of victorious Communist parties. Why did these parties first take the road of cultural leveling, and why did they then abandon this road in favor of the technocratic road? In the Chinese case, I stress the importance of ideas in answering the first question, and the importance of interests in answering the second.

EXPLAINING CLASS LEVELING

Once the CCP had taken the technocratic road, it presented its earlier forays into cultural leveling as a deviation from socialist principles. The CPSU also disavowed its early radicalism, and both parties ended up presenting the Communist mission as fundamentally technocratic. While these reinterpretations gut the radical elements from Marxist doctrine, it is true that technocratic tendencies existed in both countries from the first days of Communist power. In China, the CCP made pragmatic compromises with the educated elite, it built a highly meritocratic education system, and technocratic ideas flourished at Tsinghua and other universities. The same was true, albeit to a lesser extent, in the Soviet Union. These early technocratic tendencies, however, always existed uneasily within parties that were generally hostile to them, and in both countries they were overwhelmed by class-leveling impulses. Moreover, the radical goals and the class-struggle methods of cultural leveling in both the Soviet Union and China were derived directly from Communist ideology.

In explaining the early hostility of the CCP to the educated elite and technocratic policies, I have stressed the peasant origin of most of the party's cadres. Scholars of the Soviet Union have, in a similar fashion, pointed to the working-class origin of most Bolshevik Party members in explaining the party's early hostility to bourgeois experts. These explanations are accurate, but they point to an intermediate mechanism, not the original impetus. The CCP and the CPSU were both founded by intellectuals who—because of their interpretation of Marxism—intentionally based their movements among the lower classes and made great efforts to elevate members of humble origin into leadership positions. After taking power, they maintained this orientation, promoting workers and peasants and discriminating against intellectuals. The justification for this class bias was that lower-class cadres were the most trustworthy bearers of the Communist mission because they had less vested interest in preserving class privileges. And this was true, at least in the case of cultural capital. In sum, the CCP and the CPSU were not hostile to technocracy because their cadres were recruited from the poorly educated classes; rather, their cadres were recruited from the poorly educated classes because the parties' guiding ideology was hostile to technocracy.

I have also explained cultural class leveling, and especially efforts to redistribute educational opportunities, as a manifestation of competition between new and old elites. Scholars of the Soviet Union have made similar arguments, highlighting the CPSU's determination to replace White experts with Red experts. These arguments are sound, but if we interpret Communist efforts to redistribute cultural capital *simply* as a means to train a more politically reliable corps of specialists, we miss much of their content. In both the Soviet Union and China, ideologically driven parties experimented with a broad range of educational policies designed to eliminate the distinction between mental and manual labor. In the Soviet Union, because redistribution efforts faded quickly after a new corps of Red experts was in place, it is easy to view early radical policies through the narrow lens of inter-elite competition. This is not justified in the Soviet case, however, and still less so in the Chinese case, where cultural redistribution continued much longer, and was clearly intended to prevent the consolidation of any kind of educated elite, White or Red.

EXPLAINING THE TECHNOCRATIC TURN

Ideology is less important in explaining the technocratic turn. Because proponents of technocratic policies were always vulnerable to charges of revisionism and class conciliation, these policies were easier to defend with pragmatic rather than ideological arguments. Such pragmatism, however, was underpinned by important interests. The potency of these interests increased with the convergence of new and old elites, and the growing corps of Red experts became the key constituency and the main benefactors of technocratic policies. I have developed a detailed narrative about how evolving elite interests underpinned the technocratic turn in China, and although my knowledge is not sufficient to venture a narrative of this type in the Soviet case, I suspect that similar interests were involved.

Other scholars have attributed this turn to the inexorable effects of universal laws. Nicolas Timasheff, who produced an influential interpretation of the first quarter-century of Bolshevik rule, and Richard Lowenthal, who wrote a much-cited essay comparing the trajectories of Communist power in the Soviet Union, China, and Yugoslavia, presented arguments of this type with particular eloquence.[25] Because they recognized that victorious Communist parties were driven by class-leveling ambitions, both authors were able to capture the contradictions of these societies and explain the twists and turns of Communist power in a far more compelling fashion than those who fail to take the Communist project seriously. In each of the cases they examined, Timasheff and Lowenthal argued that a revolutionary regime attempted to impose a utopian program on a resistant population, which was more interested in getting along or getting ahead than creating an egalitarian society. Individual and group interests pervaded both Timasheff's and Lowenthal's accounts, but they connected these interests to unalterable characteristics of the human species, and they interpreted individual actions as the working out of much larger forces. For Timasheff, the Great Retreat was dictated by the resilience of tradition. The Bolsheviks, he argued, encountered relentless popular resistance because they attacked the country's cultural foundations and national pride, and were trying to tear down traditional institutions that served as Russia's social fabric, including the family, the church, and the school; in order to maintain power, they were finally compelled to abandon their utopian experiments and embrace traditional institutions and ideas. Lowenthal, in contrast, argued that Communist class leveling was perpetually at odds with simultaneous efforts to develop the national economy, and that utopianism finally gave way to the requirements of modernization.

These are powerful arguments, and they are found in different forms in the works of other scholars as well. It is certainly possible that Communist goals were vanquished by human nature, the resilience of tradition, the requirements of modernization, or by a combination of the three. If so, it is also possible that the outcome was inevitable. These are questions that this book, unfortunately, cannot answer. My inquiry has been pitched at a less lofty level, examining particular sets of interests based on existing institutions. I have concentrated on those interests connected with the unequal distribution of cultural and political capital, and the institutions that facilitated reproduction of this unequal distribution. The most important of these were the school system and the Communist Party, and so I have focused my attention on conflicts surrounding these institutions and the groups that coalesced to attack and defend them. I believe the present analysis offers an adequate explanation for why class leveling was abandoned in favor of the technocratic road in China, and I expect similar analyses would go far in explaining analogous outcomes in the Soviet Union and other countries. There may be more profound reasons that made these outcomes inevitable, but as long as explanations based on privileged groups defending vested interests seem to suffice, I am reluctant to accept that the results were dictated by inexorable universal causes.

In any case, class advantages based on unequal distribution of capital—economic, political, or cultural—will continue to provoke demands for redistribution, and the resulting conflicts will continue to spur arguments denouncing or defending institutions that reproduce unequal distribution. For decades, and probably for centuries to come, proponents of both kinds of arguments will take as a key reference point the events of the Chinese Cultural Revolution, which represented the culmination of the twentieth-century Communist class-leveling experiments.

Reference Matter

Appendix 1
Tsinghua University Faculty, Production Workers, and Students, 1949–1992

Tsinghua University Faculty, Production Workers, and Students, 1949–1992

Year	Faculty[1]	Production workers and employees[2]	New student enrollment[3]		Total student enrollment[4]		Female percentage of regular student enrollment[5]
			Regular	Vocational	Regular	Vocational	
1949	323		651	48	2,257		3.7
1950	399		658		2,494	53	
1951	452		823	77	2,820	295	11.3
1952	443		1,055	925	3,007	262	
1953	588		1,850		3,381	853	
1954	683		1,808		4,990	224	
1955	822		1,933		6,392		
1956	1,227		2,340		8,647		
1957	1,230	343	1,835		9,262		19.6
1958	1,390	1,100	2,782		10,889		17.7
1959	1,575	995	2,064		11,366		17.3
1960	1,823	2,056	2,445		13,231		18.1
1961	2,005	944	1,504		12,749		18.0
1962	2,151	538	1,421		12,153		17.7
1963	2,157	502	1,631		11,596		18.4
1964	2,226	498	1,629		10,771		18.0
1965	2,475	581	1,649		10,347		
1966		626			10,347		
1967					8,135		
1968					3,400		
1969	2,549	1,042			3,265		

1970	3,327	1,432	2,236	606	2,236	606	20.2
1971	3,386	1,486			2,157	480	
1972	3,473	1,621	1,798	274	3,895	405	
1973	3,483	1,537	1,846	123	5,779	321	
1974	3,628	1,547	3,352	116	6,569	201	
1975	3,672	1,627	3,715	426	7,912	1,039	
1976	3,756	1,735	1,960	578	7,980	1,246	
1977	3,871	1,841	1,054	401	5,511	337	
1978	3,899	1,331	1,050	425	6,170	236	
1979	3,775	1,034	1,899	214	5,905		
1980	3,723	1,036	1,956	75	7,604		
1981	3,665	942	1,977		7,792		
1982	3,622	901	1,958		8,814		
1983	3,560	876	2,075		9,832		
1992	3,327	655	2,096	181	10,044	492	18.9

1. Fang and Zhang (2001, vol. 1, 489–90).

2. Fang and Zhang (2001, vol. 1, 415–17).

3. Fang and Zhang (2001, vol. 1, 222).

4. Fang and Zhang (2001, vol. 1, 216–17).

5. *QHDXYL* (1959; 1960; 1961; 1962; 1963–64; and 1964–65); Tsinhhua University (1975), Fang and Zhang (2001, vol. 1, 223). Data from 1992 only includes new enrollment.

Appendix 2
List of Interviewees

The following information about the individuals interviewed for this book is provided: Occupation and time period at the school, gender, family origin (according to official classifications), relationship to the Communist Youth League (CYL) and the Communist Party (CP), and factional sympathy, if any, during the Cultural Revolution. School names are abbreviated as follows: Tsinghua University (TU), Tsinghua University Attached Middle School (TUAMS), Tsinghua University Attached Primary School (TUAPS), and Tsinghua University Workers and Peasants Accelerated Middle School (TUWPAMS). With one exception, these are pseudonyms.

1. Wan Shaoye. TU student 1960–69. Male. White collar. Not CYL member. Jinggangshan sympathizer.
2. Zhao Zukang. TU student 1962–69. Male. White collar. CYL member.
3. Jin Huaiyuan. TU student 1962–69. Male. Less-than-good.
4. Hou Shengdong. TU student 1963–69. Male. White collar. Not CYL member. Jinggangshan sympathizer.
5. Zhao Yong. TU student 1962–69. Male. White collar. CYL member. April 14th activist.
6. Chen Xiaogang. TU student 1964–70. Male. Revolutionary cadre. CYL member. April 14th activist.
7. Li Weizhang. TU student 1964–70. Male. White collar. CYL leader. April 14th sympathizer.
8. Hua Bowen. TU student 1961–69. Male. Jinggangshan sympathizer.
9. Liu Peizhi. TU student 1963–69. Male. White collar. CYL member. Jinggangshan activist.

10. Yu Deshui. TU student 1964–70. Male. Middle peasant. CYL leader. Old Red Guard sympathizer.
11. Wang Jiahong. TU student 1963–69. Male. White collar. CYL member. April 14th sympathizer.
12. Qiu Maosheng. TU student 1964–70. Male. Poor peasant. CYL leader and party member. April 14th activist.
13. Yue Huaiyuan. TU student 1962–69. Male. Landlord. April 14th activist.
14. Ma Yaozu. TU student 1960–69. Male. Middle peasant. CYL member and class leader. April 14th activist.
15. Zhou Wenhai. TUAMS student 1963–69. Male. Capitalist. CYL member. Jinggangshan activist.
16. Liu Jinjun. TUAMS student 1963–69. Male. Revolutionary cadre. CYL leader. Old Red Guard activist.
17. Li Jingsheng. TUAMS student 1963–69. Male. Bad. Not CYL member. Jinggangshan activist.
18. Cao Ying. TUAMS student 1963–69. Female. Revolutionary cadre. CYL leader. Jinggangshan activist.
19. Liao Pingping. TUAMS student 1963–69. Female. Revolutionary cadre. Mao Zedong Thought Red Guard sympathizer.
20. Mei Tingyu. TUAMS student 1965–69. Male. White collar. Not CYL member. Old Red Guard sympathizer.
21. Zheng Heping. TUAMS student 1964–68, TU technical-training student 1975. Male. White collar. CYL member.
22. Cai Jianshe. TUAMS student 1963–69, TU student 1977–82? Male. White collar. Not CYL member. Jinggangshan activist.
23. Ding Yuqin. TUAMS student 1963–69. Female. Worker. CYL member. Jinggangshan activist.
24. Ou Yingcai. TUAMS student 1963–69. Male. White collar. Not CYL member. Jinggangshan activist.
25. Li Mengxiong. TUAMS student 1963–69. Male. Capitalist. Not CYL member. Jinggangshan activist.
26. Song Zhendong. TUAMS student 1965–69. Male. Revolutionary cadre. CYL member. Mao Zedong Thought Red Guard sympathizer.
27. Sun Qing. TUAMS student 1965–69. Female. White collar. Not CYL member.

28. Lu Jianxin. TUAMS student 1965–69. Female. Revolutionary cadre. CYL member. Mao Zedong Thought Red Guard sympathizer.
29. Li Huan. TUAMS student 1965–69. Female. White collar. Not CYL member. Mao Zedong Thought Red Guard sympathizer. April 14 activist.
30. Wei Jieming. TU Kindergarten and TUAPS student 1974–81, TUAMS student 1981–87, TU student 1987–96. Male. White collar. CYL member (TU).
31. Ding Yi. TUAMS student 1969–75, TU student 1977–82. Male. White collar. Party member.
32. Gao Yizhi. TUAMS student 1971–75, TUAMS teacher 1975–78. Male. White collar. Not CYL member.
33. Liu Wenqing. TUAPS student 1971–77, TUAMS student 1977–81, TU student 1981–86. Male. White collar. CYL member.
34. Zhang Yongyi. TU student, TU cadre, TUAMS cadre 1995–present. Male.
35. Chen Ruowen. TUAMS student 1968–71, TUAMS cadre 1980–present. Male. White collar.
36. Zhang Guiying. TUAMS student 1956–60, TUAMS cadre 1960–present. Female. Worker. CYL leader and party member.
37. Yuan Jieqiong. TUWPAMS teacher/cadre 1950–58, TUAMS teacher/cadre 1960–? Female. White collar. CYL leader and party member.
38. Wang Zhengsheng. TUAMS teacher 1949–? Male. White collar.
39. Liang Juncheng. TU student 1949–52, TU teacher 1953–60, TUAMS teacher 1960–88. Male. CYL leader and party member.
40. Zhang Dingzhong. TUAMS teacher 1961–? Male. CYL leader and party member.
41. Dai Yingzhi. TU teacher 1955–60, TUAMS teacher 1961–? Female. Small capitalist.
42. He Xianlong. TU student 1949–52, TUWPAMS teacher 1952–57, TUAMS cadre 1957–69. Male. White collar. CYL and party leader.
43. Xue Limin. TUAMS student 1960–68, TU student 1978–82, TUAMS teacher 1982–? Female.
44. Zhao Songling. TU student 1936–40, TU teacher 1951. Female. White collar.
45. Yue Changlin. TU teacher 1953–86. Male. White collar. Not party member.

46. Wei Xuecheng. TU student 1953–58, TU teacher 1958–? Male. Rich peasant. Jinggangshan sympathizer.
47. Wei Jialing. TU student 1958–64, TU teacher 1964–? Female. Nationalist official. CYL leader and party member.
48. Yang Yutian. TU student 1956–59, TU teacher 1959–? Male. Peasant. April 14th sympathizer.
49. Liang Yousheng. TU student 1953–58, TU teacher 1958–? Male. Nationalist official.
50. Zhang Cuiying. TU student 1970–73, TU staff 1973–present. Female. Peasant. CYL leader.
51. Zhuang Dingqian. TU student 1952–53, TU teacher 1953–present. Male. White collar. CYL leader and party member.
52. Chang Zhenqing. TU student 1965–70, TU teacher 1970–present. Male. Small business. April 14th sympathizer.
53. Yue Xiuyun. TU student 1972–75, TU teacher/staff 1975–present. Female. Peasant.
54. Zhao Xianlu. TU student 1965–70, TU teacher 1970–present. Male. Jinggangshan sympathizer.
55. Fang Xueying. TU student 1972–75, TU teacher 1975–present. Female. Revolutionary cadre.
56. Xiong Minquan. TU student 1948–51, TU cadre 1951–84. Male. White collar. CYL and party leader.
57. Tong Yukun. TU student 1946–51, TU teacher/cadre 1951–84. Male. CYL and party leader.
58. Tong Xiaoling. TUAMS student 1968–70, TU staff 1970–77? Female. Revolutionary cadre.
59. Zhang Youming. TUAMS student 1961–64, TU student 1965–70, TU staff 1970–? Male. Worker. CYL leader. Jinggangshan activist.
60. Wang Xingmin. TU student 1965–70, TU staff/cadre 1970–present. Male. Peasant. CYL member. Jinggangshan sympathizer.
61. Zhao Heping. TU student 1975–81. Male. White collar. CYL member.
62. Fang Zhenzhong. TU student 1965–70. Male. Worker. CYL leader. April 14th activist.
63. Mei Xuesi. TU student 1965–70. Male. White collar. CYL member.
64. Yue Changling. TU student 1979–84. Male. White collar. CYL member.
65. Wu Xianjie. TU student 1978–83. Male. CYL member. White collar.

66. Han Lingzhi. TUAMS teacher 1976–77, TU student 1978–82. Female. White collar. CYL and party member.
67. Liang Yin. TU student 1982–84. Female. White collar. CYL member.
68. Lin Juan. TU student 1983–87. Male. Landlord. CYL member.
69. He Yi'ning. TU student 1965–70. Male. Capitalist. Not CYL member. Jinggangshan sympathizer.
70. Fei Jie. TU student 1977–82. Female. White collar. CYL member.
71. Ding Xuan. TU student 1977–82. Male.
72. Zhao Junying. TU student 1959–65. Male. Capitalist. CYL member.
73. Ke Ming. TU student 1965–70. Male. Revolutionary cadre. CYL leader and party member. April 14th activist.
74. Kuai Dafu. TU student 1963–69. Male. Poor peasant. CYL leader. Jinggangshan activist.
75. Li Guangyou. TU student 1964–70. Male. Worker. CYL member. Jinggangshan activist.
76. Lai Jiahua. TU student 1948–52, TU teacher/cadre 1952–? Male. Capitalist. Party leader. Jinggangshan sympathizer.
77. Hong Chengqian. TU student 1954–58, TU cadre 1958–present. Male. Small business. Party leader.
78. Cheng Yuhuai. TU student 1956–61, TU teacher/cadre 1961–present. Male. Middle peasant. Party leader. April 14th sympathizer.
79. Luo Xiancheng. TU student 1970–74, TU teacher 1974–present. Male. Peasant. Party member.
80. Zhu Yongde. TU student 1960–68, TU employee 1968–present. Male. White collar? CYL member. April 14th activist.
81. Chen Jinshui. TU worker 1952–? Male. Worker. Party member. Jinggangshan sympathizer.
82. Zhu Youxian. TU student 1970–74, TU teacher 1974–present. Male. Poor peasant. Party member.
83. Lin Jitang. TU student 1965–70, TU cadre/teacher 1970–present. Male. Middle peasant. Jinggangshan activist.
84. Wang Hang. TU student 1972–75. Male. Revolutionary cadre. CYL member.
85. Lin Hongyi. TU student 1963–69. Male. Peasant. CYL leader. Jinggangshan activist.

86. Yuan Zheng. TUAPS student 1960–65, TUAMS student 1965–69, TU graduate student 1978–81, TU teacher 1981–present. Male. White collar.
87. Sun Shengqian. TU student 1961–68. Male. Peasant. April 14th sympathizer.
88. Zuo Chunshan. TU student 1974–77. Male. Lower-middle peasant. Party leader.
89. Luo Jinchu. TU student 1973–77, TU teacher and staff 1977–present. Male. Revolutionary cadre.
90. Qin Yucheng. TUAMS student 1960–66, TU student 1978–82, TU teacher 1982–present. Male. White collar. CYL leader. Jinggangshan sympathizer.
91. Long Jiancheng. TU foreign student 1975–77. Male.
92. Mai Qingwen. TU student 1951–55, TU teacher/cadre, 1955–83. Male. Capitalist. Party leader. 414 activist.
93. Lu Baolan. TU student 1973–77, TU teacher 1977–present. Female. Poor peasant.
94. Luo Yaozong. TU student 1972–75, TU teacher 1975–present. Male. Worker.
95. Li Hongjun. Primary school student in village near Tsinghua 1970s. Male. Peasant.
96. Lu Shenping. TU student 1965–70. Male. White collar. CYL member.
97. Chen Zhiming. TU student 1964–69. Male. Worker. CYL member.
98. Zhen Xiaogang. TU student 1964–69. Male. Poor peasant. CYL member.

Abbreviations

HQ	*Hongqi* (Red flag)
JGS	*Jinggangshan* (Published by Tsinghua University Jinggangshan Regiment) (1966–68) (Reproduced in Yuan Zhou, editor, *A New Collection of Red Guard Publications, Part 1: Newspapers.* 1999. Oakton, VA: Center for Chinese Research Materials).
JGSB	*Jinggangshan bao* (Jinggangshan news) (Published by Tsinghua University Jinggangshan-April 14th) (1967–68) (Reproduced in Yuan Zhou, editor, *A New Collection of Red Guard Publications, Part 1: Newspapers.* 1999. Oakton, VA: Center for Chinese Research Materials).
QHDXYL	*Qinghua Daxue yilan* (Tsinghua University yearbook).
QHGB	*Qinghua gongbao* (Tsinghua bulletin) (1954–66).
QHXYTX	*Qinghua xiaoyou tongxun* (Tsinghua alumni bulletin) (1980–present).
QHZB	*Qinghua zhanbao* (Tsinghua battle report) (1970–77).
RMRB	*Renmin ribao* (People's daily).
XQH	*Xin Qinghua* (New Tsinghua) (1952–66; 1978–present).
XQH-GQTZK	*Xin Qinghua gongqingtuan zhuankan* (New Tsinghua—Communist Youth League edition).
XQH-JSZK	*Xin Qinghua jiaoshi zhuankan* (New Tsinghua—teachers' edition).
XQH-ZGZK	*Xin Qinghua zhigong zhuankan* (New Tsinghua—employees' edition).
ZGQNB	*Zhongguo qingnian bao* (China youth news).

Notes

INTRODUCTION

1. Saint-Simon's followers became stronger advocates of public property than he had been. For interpretations of the ideas of Saint-Simon and his followers, see Carlisle (1987); MacIver (1922); and Manuel (1956).

2. Elimination of the division between mental and manual labor is a basic theme in Marx's works; for one instance, see Marx (1978, 531).

3. Other technocratic accounts from this period include Bayliss (1974); Bell (1973); Galbraith (1967); Gouldner (1979); and Ludz (1972). Previous efforts to explain the rise of a new dominant class in countries ruled by Communist parties had generally focused on political power. In Milovan Djilas's 1957 book, *The New Class*, which cemented the term in the popular imagination, he described the progenitors of this class as a band of proletarian revolutionaries who destroyed the existing elite classes and leveled all competing foundations of social power; the ruling Communist party and its state bureaucracy—political power—became the sole hierarchy of social differentiation. Konrad and Szelenyi disputed this notion; although they recognized the political foundations of the New Class, they were more interested in its cultural foundations, which to them seemed to portend a technocratic future not only for socialist societies, but perhaps for the entire world.

4. The terminology of political and cultural capital was not used by Konrad and Szelenyi, but was adopted by Szelenyi in subsequent works on this topic. A few years after *Intellectuals on the Road to Class Power* was published in 1979, Szelenyi (1986) conceded that the New Class project—making knowledge the main basis of class power—had been at least temporarily obstructed in Eastern Europe by bureaucratic elites' efforts to retain their political monopoly. Up until 1989, however, Szelenyi maintained that the project was more likely to succeed in the East than in the West because planning was more congenial to such a project than the market (Szelenyi and Martin 1988). Since 1989, however, he and his colleagues have argued that the New Class project survived the demise of communism and has been able to flourish under capitalist banners (Eyal, Szelenyi, and Townsley 1998).

5. Konrad and Szelenyi (1979, 184–92, 203).
6. See Chapter 10.
7. Burawoy (1998, 5).
8. Three monograph-length studies have examined the technocratic transformation of the CCP: Lee (1991); Li (2001); and Zang (2004). A number of recent quantitative analyses have documented the persistence of dual political and technical career tracks, and have confirmed that Chinese officials increasingly must have academic credentials, but still must have political credentials; see Bian, Shu, and Logan (2001); Dickson and Rublee (2000); Walder, Li, and Treiman (2000); and Zang (2001). In addition, a number of scholars have analyzed the characteristics of members of CCP central leadership bodies (Li and White 1988; Li and White 1990; Li and White 1991; Li and White 1998; Li and White 2003; North and Pool 1966; Scalapino 1972; Zang 1993); of officials in individual cities (Chamberlain 1972; Kau 1969; Lieberthal 1980; Vogel 1967; Vogel 1969; Wang 1995; White 1984); and of particular groups of Communist leaders (Israel and Klein 1976).
9. See, for example, Chen (1960); Goldman (1967); Goldman (1970); Goldman (1981); Liu (1990); MacFarquhar (1960); Mu (1963); and essays collected in Goldman, Cheek, and Hamrin (1987); Gu and Goldman (2004); and Hamrin and Cheek (1986).
10. Important works on post-1949 Chinese education policy include Chen (1981); Cleverly (1985); Cui (1993); Han (2000); Hayhoe (1996); Pepper (1996); Taylor (1981); Unger (1982); White (1981); Zhou (1999); and Zhu (2000). In addition to discussing conventional educational goals, a number of these authors have considered the role of the education system as a mechanism of class differentiation, and several quantitative analyses have gauged the impact of changing policies on inequality in educational attainment; see Deng and Treiman (1997); Hannum (1999); Hannum and Xie (1994); Knight and Shi (1996); and Liu (1999). Among the studies that have analyzed the structure and function of the party and state system are Barnett (1967); Harding (1981); Lewis (1963); Lieberthal (1995); Schurmann (1968); Walder (1986); Whyte (1974a); and Zheng (1997). Among the studies that have specifically examined the party and Youth League recruitment apparatus are Bian, Shu, and Logan (2001); Funnel (1970); Leader (1974); and Shirk (1982).
11. See Chan, Rosen, and Unger (1980); Lee (1978); Liu (1986–87); Rosen (1982); Wang (1995); White (1976); Yin (1997a); and Xu (1996).
12. For an overview of his tripartite framework, see Bourdieu (1986); for more elaborate discussions of class differentiation in the cultural field in France, see Bourdieu (1984) and Bourdieu (1989). Bourdieu's definition of class is broader than that employed by Weber, who limited class to market position (Weber 1978, 926–40). It is also broader than that employed by many Marxists, who define class largely as property ownership, but it is compatible with Marx's broader definition of class as position in the relations of production. In discussing precapitalist societies, Marx

emphasized that a person's class position was determined largely by his or her social position and possession of skills, while private ownership of property played a less important role (Marx 1973, 491–502).

13. Szelenyi's adaptations can be found in Konrad and Szelenyi (1991) and Eyal, Szelenyi, and Townsley (1998). Sun (2002) used this tripartite framework to develop an insightful analysis of class transformation in China.

14. Although terminology varies, there is wide agreement among scholars that in the absence of private property the principal mechanisms of class differentiation in socialist societies were political and cultural. See, for instance, Bell (1973); Gouldner (1979); Inkeles (1966); Lane (1982); Parkin (1971); and Wright (1994).

15. The classic treatise on human capital is Becker (1964).

16. Bourdieu (1986, 248–49).

17. When Bourdieu (1998) briefly discussed the class structure of socialist East Germany, he stressed the central importance of party membership, which he treated as a political form of social capital.

CHAPTER 1

1. Liu later became one of the founders of the People's Navy and served as its deputy commander. Zhou was transferred to Peking University as part of the reorganization of higher education in 1952; he would eventually become president of the university and serve as honorary chairman of the National People's Congress. Biographies can be found at: http://www.library.hn.cn/difangwx/hxrw/xdrw/jfj/liudaosheng.htm and http://www.cast.org.cn/n435777/n435799/n676835/n677237/20641.html.

2. North and Pool (1966, 389).

3. Townsend (1970, 303); and Lee (1991, 45).

4. Many future Communist leaders tested into rural middle schools, often tuition-free teachers' schools, where they were recruited by underground party organizers (Cong 2007).

5. Hinton (1966); and Friedman, Pickowicz, and Selden (1991).

6. See Brugger (1976); Gardner (1969); Lieberthal (1980); and Vogel (1969).

7. Bo quoted in Zhu (2000, 1494).

8. Chamberlain (1972) and Wang (1995) documented the situation of "dual elites" in Tianjin, Guangzhou, Shanghai, and Wuhan.

9. Li (2001).

10. The United States and several European powers compelled the Qing dynasty to pay indemnities after they invaded China to suppress the anti-Western Boxer Rebellion in 1900.

11. See Table 3.2. This survey used the CCP's class categories.

12. Yuan Yongxi, who served as Tsinghua's party secretary until he was displaced by Jiang in 1956, was also highly educated.

13. Fang and Zhang (2001, 25–26); and Li (1994, 10).

14. See Whyte (1974a).

15. Mann (1986).

16. Estimates of the proportion of the population aged fifteen to twenty-five that belonged to the Youth League are cited in Leader (1974, 701). In 1965 there were approximately eighteen million party members (Lee 1991, 17) out of an adult population of about 376 million (Chinese Academy of Social Sciences Population Research Center 1985, 602–3; data is from the 1964 Census). Party and league members made up a higher proportion of the urban population and in some factories one-fifth of employees belonged to the party.

17. Liu and Fang (1998, 536).

18. For descriptions of the recruitment process and criteria, see Montaperto (1972); and Shirk (1982).

19. This proportion includes students who had left the league and joined the party. Data was provided for students who graduated in the winter and summer of 1963. See *QHGB* (March 16, 1963; May 11, 1963).

20. *QHGB* (March 16, 1963; May 11, 1963).

21. See Table 3.5.

22. For analyses of the continuing importance of political credentials in the administrative track, see Walder (1995); and Walder, Li, and Treiman (2000).

23. Interviewee 63.

24. Shirk (1982) wrote that the CCP strived to select young people of good moral character because, like other ideological revolutionary movements, it was seeking to create a "virtuocracy."

25. Interviewee 58.

26. Interviewee 63.

27. Liu (1967).

28. The class origin system grew out of the Land Reform campaign, in which each rural family was assigned a class designation as part of investigations that preceded property redistribution; the system was later extended to urban areas.

29. Although class origin and political background were formally differentiated (official forms typically provided different places for each), in the popular consciousness and in practice the two categories were often conflated. This practical conflation has often been repeated in scholarly discourse.

30. Krauss (1981, 20–26); Wang (1995, 25–33); White (1976, 2); and White (1984, 143–44).

31. Advantages derived from peasant and proletarian class designations should be distinguished from those derived from association with the Communist Party. Both were political advantages, but peasant family origin was hardly a predictor of success, while association with the party was.

32. Interviewee 51.

33. Interviewee 92.

34. As Walder (1986) pointed out, limited mobility and dependence on goods and services distributed by work units enhanced the power of work unit leaders over members.

35. See Teiwes (1976). Hinton (1966) provides a fascinating ethnographic account of a party work team organizing villagers to criticize local Communist cadres during a 1947 party rectification campaign.

36. Chen (1960); Goldman (1967); MacFarquhar (1960); and MacFarquhar (1974).

37. Mao quoted in Zhou (1999, 22).

38. Mao quoted in Liu and Fang (1998, 561).

39. Mao quoted in *JGS* (Sept. 6, 1967, 4). Dongchang'an Street refers to Zhongnanhai, the party's headquarters in Beijing. The city of Shijiazhuang was captured by the CCP in a key civil war battle and became a launching pad for the assault on Beijing. The small city of Yan'an served as the Communist headquarters during the anti-Japanese War.

40. Interviewee 45.

41. *New Tsinghua* Editorial Committee (1957, 186).

42. *New Tsinghua* Editorial Committee (1957, 183–84).

43. Hinton (1972, 36); and Li (1994).

44. *XQH* (May 22, 1957, 3).

45. *New Tsinghua* Editorial Committee (1957, 187).

46. *XQH* (May 18, 1957, 4).

47. *New Tsinghua* Editorial Committee (1957, 203–6).

48. *New Tsinghua* Editorial Committee (1957, 77–78).

49. *XQH* (June 15, 1957, 1; June 22, 1957, 1; June 24, 1957, 4; Feb. 11, 1958, 3).

50. *XQH* (July 4, 1957, 1).

51. For accounts of the Anti-Rightist movement, see Chen (1960); Goldman (1967); and MacFarquhar (1974). The 403 Tsinghua students denounced as Rightists made up 4.4 percent of the student body. Assuming the great majority of the 168 employees criticized as Rightists were teachers, over 10 percent of the university faculty may have been so labeled. See Liu et al. (1987, 70); and Fang and Zhang (2001, vol. 1, 216, 490, 521, 525).

52. *XQH* (June 6, 1980, 1).

53. *JGS* (Sept. 6, 1967, 4).

54. Interviewee 48.

55. Interviewee 46.

56. See, for instance, Chang and Halliday (2005, 416–21).

57. MacFarquhar (1974).

58. Goldman (1967).

CHAPTER 2

1. *New Tsinghua* Editorial Committee (1957, vol. 2, 120).
2. Andreas (2004, 18).
3. See Chapter 3.
4. See the concluding chapter of this book.
5. Chen (1981); Cleverly (1985); Hayhoe (1996); Pepper (1996); and Unger (1982).
6. Andreas (2004, 18); and Jiang (1998, 853).
7. Hayhoe (1996).
8. Israel (1982–83, vi); and Li (2001).
9. Interviewee 57.
10. Naughton (2007, 56). See also Riskin (1991).
11. State Education Commission (1984, 56–58).
12. Cadre positions could be administrative or technical; the cadre title distinguished these positions from those of workers.
13. Mao (1977, 71).
14. MacFarquhar (1974).
15. Mao (1974, 116).
16. Interviewee 48.
17. The workers' school was established in 1955. See Fang and Zhang (2001, vol. 1, 301–2), and *XQH* (Oct. 15, 1958, 2).
18. Liu et al. (1987, 27, 81); Wan (1987, 8–20); and *XQH* (June 23, 1958, 2).
19. See Table 3.4.
20. Interviewee 46.
21. Interviewee 48.
22. In 1965, Jiang was named minister of higher education, a new post.
23. Deng Xiaoping presided over the conference that adopted these guidelines for higher education, for which Tsinghua served as a model. Tsinghua officials Li Shouci, He Dongchang, Ai Zhisheng, and Gao Jingde played important roles in the drafting process.
24. Keypoint schools are discussed in depth by Pepper (1996); Unger (1982); and Yuan (1999).
25. Wan (1987, 22–67).
26. *JGS* (Dec. 21, 1967, 3).
27. Jiang (1998, 810–11).
28. *XQH-JSZK* (Oct. 11, 1962, 3).
29. Interviewee 48.
30. Mao (1974).
31. Interviewee 49.
32. *XQH* (June 26, 1965, 4; Dec. 31, 1965, 1; April 24, 1966, 1).
33. Rural-oriented education programs had difficulty surviving next to regular

academic schools, where the curricula were oriented to the examination system. Andreas (2004) and Pepper (1996) examined the failure of agriculture middle schools before the Cultural Revolution.

34. Jiang (1998, 854–55).

CHAPTER 3

1. *XQH* (March 22, 1958, 1).

2. Schurmann (1968, 129) cited a figure of 4,448,080 party members as of October 1949. Gu (1984, 141) cited the total number of graduates from Chinese institutions of higher education from the time the first one was established at the end of the nineteenth century until 1949. Some of these graduates had certainly already died by 1949.

3. Andreas (2004, 18).

4. North and Pool (1966, 376–82). Other Central Committee members had been sent to the Soviet Union for training.

5. In 1936, Tsinghua's underground party organization had forty-two members; in 1948, two hundred students—10 percent of the university student body—were party members. During both periods, Tsinghua had the strongest CCP organization among Beijing universities (Fang and Zhang 2001, 790, 796). For an account of the December 9 movement at Tsinghua and other schools and the political careers of students who joined the CCP, see Israel and Klein (1976). Li (2001, 108–9) presented biographical data on thirteen important CCP leaders who were active in the December 9 movement at Tsinghua or who joined the party while students at the university in the late 1940s.

6. Lee (1991, 45).

7. Bourdieu's metaphor is imprecise because conversion does not necessarily require an exchange or transformation of capital.

8. Taylor (1981, 132).

9. It was rare for students to win admission if they had not excelled in the examination competition. One case of parental intervention, in which the son of a top general, He Pengfei, was admitted to Tsinghua's attached middle school for a year after low exam scores prevented him from directly entering Tsinghua University, was famously exposed during the Cultural Revolution. The very notoriety of the case seems to indicate this kind of violation of admissions procedures was not common. This was the consensus among teachers and students I consulted.

10. Rosen (1979); and Unger (1982).

11. Jiang (1998, 597).

12. Jiang (1998, 586–88, 594–609).

13. *QHZB* (Oct. 12, 1970, 4).

14. Jiang (1998, 814).

15. Tang (1998).

16. Li (2001, 94).
17. Fang and Zhang (2001, vol. 1, 817).
18. Interviewee 1.
19. Interviewee 22.
20. Interviewee 46.
21. *ZGQNB* (Jan. 4, 1958).
22. Interviewee 21.
23. For an account of how the CCP groomed workers for leadership positions, see Harper (1971).
24. *Statistical Work Dispatch* Materials Office (1957, 89).
25. Lee (1991, 295).
26. Jiang (1957, 12).

CHAPTER 4

1. The Peking University teachers, led by Nie Yuanzi, received encouragement from the wife of Kang Sheng, a member of the CCRSG, which Mao had just established to lead the movement (Harding 1991, 134–35).
2. Recent books that stress Mao's pursuit of personal power in explaining the origins of the Cultural Revolution include Chang and Halliday (2005); and Huang (2000).
3. See, for instance, Dittmer (1987); Ahn (1974); Whyte (1974b); Lee (1978); Meisner (1982); Schapiro and Lewis (1969); Schwartz (1968); and Tsou (1969).
4. This division of labor was formalized in 1956: Mao, in the "second front," was responsible for "questions of principle," while other top leaders, in the "first front," were charged with handling daily affairs (Huang 2000, 13; and MacFarquhar 1974, 152–53).
5. Elsewhere, I have attributed the effectiveness of the Cultural Revolution rebel movement in undermining bureaucratic authority to its charismatic character (Andreas 2007).
6. Mao (1996, vol. 2, 265–66).
7. *RMRB* (Aug. 9, 1966).
8. For accounts of work team methods during the Land Reform movement, see Friedman, Pickowicz, and Selden (1991); and Hinton (1966). For accounts of the Socialist Education movement, see Chan, Madsen, and Unger (1984); Endicott (1988); and Yue and Wakeman (1985).
9. Mao (1996, vol. 2, 265–66).
10. From a speech delivered by Mao in February 1967, quoted in *Peking Review* (Feb. 13, 1976, 5).
11. For accounts of the role of the CCRSG and the background of its members, see Dittmer (1987, 80); and Barnouin and Yu (1993, 44–55).

12. Bu (1998).
13. Zhong (1996).
14. Zheng (1992a, 1992b).
15. Zhong (1996) and Interviewees 18 and 20.
16. Mao (1988).
17. Interviewee 19.
18. Song (2004).
19. Tsinghua University Attached Middle School Red Guards (1966).
20. Zhong (1996) and Interviewees 17, 18, and 23. Rosen (1979, 186) noted that two or more competing Red Guard organizations emerged in many Beijing middle schools in August 1966, sometimes reflecting conflict between revolutionary cadres' children and children of workers and peasants. At Tsinghua's attached middle school, however, one Red Guard organization dominated the school until an opposition faction, composed mainly of children of intellectuals, emerged in late 1966.
21. Tsinghua University Attached Middle School Red Guards (1966).
22. Interviewee 26.
23. Tsinghua University Attached Middle School Red Guards Senior Middle School Class Number 655 Leadership Small Group (1996 [1966]).
24. Wang (1996).
25. Reprinted in Song and Sun (1996, 108).
26. Interviewee 16.
27. Interviewee 22.
28. Tsinghua University Attached Middle School Jinggangshan Regiment (1967).
29. Interviewee 22.
30. Interviewee 18.
31. Yu was executed in 1970 during a wave of repression against rebel activists. His execution was one of the most egregious cases of suppression of political expression during the Cultural Revolution decade, and he became a celebrated martyr among Chinese dissidents (Song 2004).
32. For an English translation, see White (1976, 71–93). The original is included in Song and Sun (1996, 120–40).
33. Chan (1985, 233); Rosen (1979, 196–204); and Yin (1997b).
34. Interviewee 17.
35. Chan, Rosen, and Unger (1980); and Rosen (1982).

CHAPTER 5

1. For a dramatic account of the armed conflict on the Tsinghua campus, see Hinton (1972).
2. The terms "radical" and "moderate" are widely used by former activists and scholars today. At the time, the radicals denounced their opponents as "conservatives," while the moderates denounced their opponents as "ultra-Leftists."

3. Because Kuai's father remained in the village, he remained a peasant, rather than a state cadre. Like other villagers, he did not have a state salary, but instead earned work points, entitling his family to a share of the production brigade's harvest.

4. *XQH-GQTZK* (Dec. 17, 1964, 1).

5. Interviewee 74.

6. Mao quoted in Foreign Languages Department Revolutionary Committee (1968, 94).

7. Kuai (1966, 4).

8. Lupher (1996) identified the Cultural Revolution as an instance of "top-and-bottom-versus-the-middle strategy of power restructuring" described by Max Weber, in which the power of an elite group is weakened by the concerted action of a central ruler and social groups at lower echelons of the social hierarchy. At numerous moments in Chinese history, Lupher pointed out, both central rulers and peasants gained by attacking "evil gentry" and "corrupt officials" who competed with central power.

9. Interviewee 73.

10. For instance, in an interview with Snow (1971, 174–75), Mao expressed dismay at the exalted titles used to identify him.

11. In some places, military representatives were able to quickly take charge; in others, including Tsinghua University, they were never able to establish their authority.

12. *HQ* No. 5 (March 1967).

13. Lee (1978, 168–74).

14. Interviewee 73.

15. Kuai's organization was henceforth popularly known as the "Regiment." I use "Jinggangshan" to refer to the radical faction so as not to unduly burden the reader with organizational names.

16. Hinton (1972), Tang (2003), and Zheng (2006) have recounted the twists and turns of the factional conflict at Tsinghua. Hinton was persuaded by the official position in 1972 that there were no significant differences between the radical and moderate factions. Tang (1998) described the political differences between the two factions much as I have.

17. Kuai (1966, 61).

18. Interviewee 74.

19. *JGS* (July 5, 1968).

20. *JGS* (May 13, 1967).

21. *JGS* (April 5, 1967).

22. Liu (1967).

23. Dittmer (1998).

24. Mao quoted in Dittmer (1998, 317).

25. *JGS* (April 18, 1967).

26. *JGS* (April 18, 1967).
27. Interviewee 9.
28. Tsinghua Jinggangshan Regiment (1967b).
29. Interviewee 9.
30. *JGS* (Nov. 17, 1967; Nov. 24, 1967).
31. *JGS* (Nov. 9, 1967).
32. *JGS* (May 1, 1967; May 8, 1967; May 13, 1967; Nov. 17, 1967); and Tang (1996, 58).
33. Interviewee 74.
34. *JGSB* (Dec. 1, 1967).
35. *JGSB* (July 5, 1967).
36. Tang (1998); *JGSB* (Dec. 14, 1967).
37. Tsinghua Jinggangshan United General Headquarters April 14 Cadre Office (1967, vol. 1, 2).
38. Interviewee 80.
39. Interviewee 14.
40. *JGS* (Oct. 19, 1967).
41. *JGS* (Oct. 19, 1967; Dec. 12, 1967).
42. Interviewee 75.
43. *JGSB* (Nov. 11, 1967; Dec. 14, 1967).
44. *JGSB* (Dec. 14, 1967).
45. Shen (2004), 193–94.
46. Shen (2004, 7–9).
47. Shen (2004, 7–8).
48. Interviewee 74.
49. For a more detailed analysis of the leadership of Tsinghua's student factions, see Andreas (2002).
50. See Shen (2004, 115).
51. Interviewee 74.
52. Shen (2004, 119–20).
53. Zhou (2006, 69). Zhou, who coauthored the famous moderate manifesto, "April 14 Will Win!," was the son of a Nationalist military officer.
54. Interviewee 12.
55. *JGSB* (Sept. 5, 1967); Tsinghua Jinggangshan United General Headquarters April 14 Cadre Office (1967, vol. 1, 5); and Tang (1996, 57).
56. *JGS* (March 23, 1968); and Tsinghua Jinggangshan United General Headquarters April 14 Cadre Office (1967, vol. 1, 4). The role of the Red Teachers Union was discussed at length by Tang (2003); some of his interpretations have been contested by Tao Dejian, a leader of the organization, and her husband Tao Shilong (http://personal.nbnet.nb.ca/stao/tdj.htm).
57. See, for instance, Lee (1978).

58. *JGSB* (July 5, 1967).
59. Interviewee 7.
60. Interviewee 11.
61. Chan, Rosen, and Unger (1980); Rosen (1982); and Unger (1982).
62. In challenging the competing elites model, Walder (2002, 2006) pointed out that students in Beijing universities did not split along family origin lines, a finding that echoed Rosen's (1979) own findings, and is consistent with the present analysis of Tsinghua University.
63. Tang (1996, 49); and Song and Sun (1996, 365).

CHAPTER 6

1. For detailed accounts of this confrontation, see Hinton (1972); and Tang (2003).
2. Mao quoted in Joint Publications Research Service (1974, 493).
3. Mao quoted in *Zhongfa* (1976, No. 4), reproduced at: http://www.zggr.org.
4. The radical-conservative factional polarization took shape more clearly after Lin Biao's fiery death in 1971. Until then, the situation was complicated by the power of Lin and other military officers, who did not fit into either camp. For one interpretation of the factional conflict that preceded and followed Lin's death, see MacFarquhar (1991).
5. The terms "radical" and "conservative" are appropriate in relation to the Cultural Revolution program, which set the agenda for the 1966–76 decade because the radicals championed this program and the conservatives opposed it. After 1976, the term "conservative" was typically used to refer to those who opposed market reforms championed by Deng Xiaoping. The terms "Left" and "Right" would have more consistent meaning, but the term "Rightist" has been indelibly associated with the targets of the 1957 Anti-Rightist movement.
6. Big Criticism Group of Beijing University and Tsinghua University (1975).
7. *QHZB* (May 10, 1976, 2).
8. *RMRB* (July 23, 1976).
9. Ma (1976, 9, 11–12).
10. Mao sent officers from Division 8431 to take charge of forming revolutionary committees in six Beijing factories, and they mobilized the workers who took over Tsinghua in July 1968. Later, these six factories along with Tsinghua and Beijing universities served as models to promote radical policies (Tang 2005).
11. See Ding (2000); and He (2001, 279, 284).
12. For a detailed account of this period, see Hinton (1972). The most harrowing moment for former student activists came during the investigation of "May 16 elements" in 1971. The investigation at Tsinghua was abruptly called off after a former student killed himself (Interviewee 52).

13. Liu and Fang (1998, 621). This number presumably includes people who took their own lives or died from health problems aggravated by the investigation.
14. He (2001, 274–75).
15. A prominent exception was former university party secretary Jiang Nanxiang, who was assigned to work in one of Tsinghua's factories (Jiang 1998, 1231).
16. Liu (1998).
17. Liu and Fang (1998, 129).
18. Interviewee 92.
19. Interviewee 3.
20. Interviewee 3.
21. Interviewee 57.
22. Interviewee 77.
23. Interviewee 55.
24. Interviewee 51.
25. Interviewee 83.
26. Interviewee 82.
27. Interviewee 55.
28. Interviewee 55.
29. Interviewee 79.
30. Interviewee 47.
31. Interviewee 48. To "put a hat on" a person meant to give someone a disparaging label such as "Rightist," "bourgeois academic authority," or "capitalist roader."
32. Interviewee 91. Long, who was from New Zealand, was part of a small contingent of foreign students who attended Tsinghua.
33. Interviewee 55.
34. Interviewee 49.
35. Interviewee 49.
36. Interviewee 81.
37. Interviewee 81.
38. Interviewee 77.
39. *RMRB* (July 27, 1975).
40. The number of party members at the university increased from 3,287 in 1965 to 7,021 in 1976, and then fell sharply, reaching a low of 3,414 in 1982 (Fang and Zhang 2001, 819).
41. Interviewee 77.
42. Interviewee 55.
43. Interviewee 55.
44. *HQ* (Jan. 1, 1976), translated in *Survey of People's Republic of China Magazines* (Jan. 23–30, 1976, 73).
45. Chi and Xie (1976, 3).
46. Yan and Gao (1996, 459).

47. Dittmer (1978, 26–60); and Meisner (1986, 418–19).

48. Radicals also gained significant administrative power in the mass media and in several municipal and provincial governments. See Chang (1979); and Dittmer (1978).

49. Han (2000).

50. For accounts of conflict over education policy during this period, see Chen (1981); Cleverly (1985); Cui (1993); Pepper (1996); Unger (1982); and Zhou (1999).

51. See Goldstein (1991); Teiwes (1984); Tsou (1995); and Zweig (1989).

52. Chan, Rosen, and Unger (1985); Forster (1990); Forster (1992b); Leijonhufvud (1990); Perry and Li (1997); Sheehan (1998); and Wang (1995).

CHAPTER 7

1. *QHZB* (March 31, 1975; July 18, 1975; May 6, 1976; May 10, 1976); and He (2001, 277–78, 287); Interviewee 47.

2. Official histories produced by Chinese education authorities typically provide detailed accounts of education before and after the Cultural Revolution decade bridged by one or two paragraphs of boilerplate text about the "ten years of turmoil" in between. Several books (Cui 1993; Zheng 1999; and Zhou 1999) have examined Cultural Revolution education policies in more detail and more fully elaborated official criticisms of these policies.

3. See, for instance, Hayhoe (1996) on universities and Unger (1982) on urban middle schools.

4. Deng and Treiman (1997); Hannum (1999); Hannum and Xie (1994); Knight and Shi (1996); Liu (1999); and Zhou, Tuma, and Moen (1998).

5. Pepper (1996).

6. Han (2000).

7. This meeting was organized by the Science and Education Group of the State Council, an ad hoc group that developed policy guidelines before the Ministry of Education was reestablished.

8. He (2001, 280).

9. Liu Bing first wrote Mao in August and then wrote again in October after receiving no response to the first letter. Several leaders of the workers' propaganda team at Tsinghua also signed the letters (Liu 1998; and Teiwes and Sun 2007).

10. Between 1965 and 1977, the number of senior middle schools in rural areas grew from 604 to 50,916, while the number of junior middle schools grew from 8,628 to 131,265 (State Education Commission 1984, 196). Many of the junior middle schools were created simply by adding grades to primary schools.

11. Because places in senior middle schools continued to be limited, it was still necessary to devise methods to select among junior middle school graduates. Methods varied by district, but usually combined recommendation and examinations.

12. Andreas (2004, 18).

13. An assessment of education quality during the Cultural Revolution decade would depend on the standards used. Cultural Revolution policies stressed practical knowledge and rejected previous education standards as scholastic. If we use students' acquisition of the content included in conventional curricula as the standard, Cultural Revolution policies led to a serious decline in quality, especially in schools that previously had enjoyed keypoint status.

14. Elsewhere I show that the elimination of college entrance examinations played an important role in facilitating the expansion of village schools and the development of rural-oriented curricula (Andreas 2004).

15. Mao quoted in Tsinghua University Workers Peoples Liberation Army Propaganda Team (1970).

16. *RMRB* (March 15, 1976).

17. The "two estimates" were contained in a key policy document issued by the National Education Work Meeting convened by the State Council from April and August 1971. Chi Qun was in charge of drafting the document, which was approved by Deputy Premier Zhang Chunqiao (He 2001, 280).

18. *QHZB* (Sept. 26, 1971, 4).

19. Interviewee 47.

20. Interviewee 60.

21. *Di yi nian tu, di er nian yang, di san nian bu ren die he niang* (*QHZB*, July 27, 1971, 3).

22. Many urban cadres were sent to live and work in rural May 7 cadre schools. These schools were named after Mao's May 7 Directive, in which he encouraged the urban population to engage in agriculture and the rural population to engage in industry.

23. For an account of life at the Liyuzhou farm, see Yue and Wakeman (1985, 251–73). Yue taught at Peking University, which also sent teachers and cadres to Liyuzhou.

24. Interviewee 49.

25. Interviewee 76.

26. Interviewee 77.

27. Interviewee 58.

28. Li and Jiang (1994, 84).

29. Interviewee 77.

30. *QHZB* (July 2, 1975, 1). The aim was to create a teaching staff that embodied a "three-in-one" combination in which teachers originated from three sources: the old teaching staff, technicians, and workers. Other departments organized similar half-work/half-study programs.

31. Tsinghua had greater resources than other schools, but school-factory collaboration was widely practiced. Teachers I interviewed in rural middle schools were

proud of the collaboration between their schools and small factories established by their communes (Andreas 2004).

32. Interviewee 51.

33. Interviewee 89.

34. Interviewee 76.

35. Whyte (1974a) argued that small group organization in China provided powerful structural incentives that encouraged a collectivist ethic.

36. Interviewee 94.

37. Interviewee 88.

38. *QHZB* (May 17, 1973; June 13, 1973).

39. Interviewee 88.

40. Interviewee 50.

41. *HQ* (May 1, 1975).

42. *QHZB* (Jan. 29, 1975, 4); *RMRB* (Dec. 30, 1975); and *China Reconstructs* (Nov. 13, 1976, 28).

43. *RMRB* (Dec. 30, 1975).

44. These schools were distinct from the May 7 cadre schools, such as the Jiangxi farm where Tsinghua cadres and teachers lived in 1969.

45. Chaoyang Agricultural University was similar to the Jiangxi Communist Labor University established during the Great Leap Forward. In his history of the Jiangxi school, Cleverly (2000) recounts the competition between the two models. The Jiangxi school was the subject of the film *Juelie* (*Breaking with Old Ideas*), produced in 1976 to promote radical education policies.

46. *QHZB* (March 31, 1975; July 18, 1975).

47. *QHZB* (March 31, 1975; July 18, 1975; May 6, 1976).

48. Interviewee 93.

49. *QHZB* (July 18, 1975).

50. State Education Commission (1984, 47, 50).

51. Zhou (1999, 225).

52. Interviewee 66.

53. This phrase from Mao's 1927 report on the peasant movement in Hunan was frequently cited in the press during the Cultural Revolution decade (Mao 1965, 29). In this report, Mao began to articulate a thesis that recurred regularly in later works: that excesses were necessary during periods of struggle and could be rectified during subsequent periods of unity.

CHAPTER 8

1. Four hundred worker-peasant-soldier students participated in twenty-two "pilot Education Revolution classes" at Tsinghua for seven months in 1969. In June 1970, the first regular worker-peasant-soldier students started arriving at the university, but the bulk of the new students did not arrive until August (He 2001, 277–78).

2. The official critique is presented well by Cui (1993) and Zhou (1999). The conflict surrounding the recommendation system is described by the above authors and by Chen (1981); Cleverly (1985); Unger (1982); and Zheng (1999). Andreas (2004) and Unger (1984) analyzed the impact of the replacement of exams with recommendation on middle school education. Working with limited available data, Bratton (1979), Pepper (1978), and White (1981) made astute observations about problems inherent in the recommendation process. Based on interviews with émigré rural teachers, Pepper (1996) produced a detailed analysis of actual recommendation practices, which she found varied greatly from place to place. Gao (1999, 107–14) recounted how recommendation worked in his own village.

3. Peking University Education Revolution Group (1972).

4. On the decline of cadres' authority and their concern about being criticized by subordinates during the late Cultural Revolution years, see Han (2000, 66–71); Perry and Li (1997, 191–92); Sheehan (1998, 139–55); and Walder (1986, 205–10).

5. Shirk (1982).

6. *QHZB* (Aug. 13, 1973). Zhang Tiesheng was arrested in 1976 and authorities denounced Zhang Chunqiao for concocting his case (Broaded 1990).

7. *RMRB* (Sept. 22, 1973).

8. Big Criticism Group of Beijing University and Tsinghua University (1975).

9. Liu and Fang (1998, 639–40). This episode was said to be inspired by celebrated writer Lu Xun's proposal to examine the examiners.

10. *RMRB* (Jan. 18, 1974, 1).

11. Jiang, Yao, Chi, and Xie (1976, 42–44).

12. Wang (2001).

13. Zhang (2000).

14. Marshal Ye Jianying was minister of defense and a key player in the factional conflicts of the 1970s. At a January 1974 Political Bureau meeting, Jiang Qing accused Ye of opening the backdoor for his own children (Wang 2001).

15. Teiwes and Sun (2007, 149–55); Wang (2001); and Zhang (2000).

16. I interviewed nineteen people who experienced the recommendation process in villages, and sixteen people who experienced the process in factories. I do not have interview data about the recommendation process in military units.

17. Interviewee 50.

18. Interviewee 82.

19. Interviewee 50.

20. Interviewee 88.

21. Interviewee 65. Wu tested into Tsinghua University after the examinations were restored in 1977.

22. Interviewee 22.

23. Pepper also reported that rural teachers she interviewed stressed the importance of labor as a criterion for recommendation "whenever mass opinion was consulted, which it often was" (Pepper 1996, 462).

24. Interviewee 93
25. Interviewee 97.
26. Interviewee 97.
27. Interviewee 95.
28. Of the three, only Lu ultimately became a worker-peasant-soldier student, entering Tsinghua in 1974. Huang was recommended by his village, but did not make the final list at the commune level. Feng was recommended by her workshop and selected by a vote of the whole factory (which chose four out of ten workshop nominees). The factory leadership then selected three of the remaining four, leaving her out. Feng thinks the factory director, who was a friend of her father (a top city official), may have been concerned that people would see approval of her recommendation as a favor to her father. Both Huang and Feng tested into college after examinations were restored in 1977.
29. Data on the family origin of students before the Cultural Revolution is presented in Table 3.4.
30. Three of the eleven Tsinghua worker-peasant-soldier students I interviewed were revolutionary cadres' children. These eleven individuals hardly constituted a random sample, as I depended on personal introductions, and most interviewees were part of a small minority who remained at the university after graduation.
31. Students who studied computer engineering at Tsinghua in 1973 and 1975 reported that several of their classmates were children of high and middle-level cadres; a number of them had been sent by the petrochemical ministry.
32. Pepper (1996, 455–65) found that nearly half of rural origin candidates recommended for admission to college or specialized middle school by the communes or state farms in which her interviewees had resided were from local cadre families. She included in this category production team leaders, most of whom were not party members and whose social and economic position differed little from other team members, but whose leadership position may have facilitated selection of their children.
33. Interviewee 82.
34. Pepper (1978) and White (1981) expressed skepticism about the actual class origin of many of the worker-peasant-soldier students, noting that everyone recommended by communes, including urban children sent down to the countryside, were officially called "village youth." The 1970 Tsinghua survey, however, recorded both family origin and work unit origin (village, factory, or military), and reported that 81 percent of entering students were of worker or poor and lower-middle peasant family origin. Students I interviewed who attended Tsinghua during this period reported that the great majority of their classmates were, indeed, from working-class and peasant families.
35. See Appendix 1. The gender imbalance was especially skewed at Tsinghua, an engineering school. In 1965, females made up 27 percent of all college students;

by 1976 the proportion had increased to 33 percent (State Education Commission 1984, 40).

36. I did not specifically seek out either female or rural origin students to interview. All three women are at Tsinghua today. And of all the women I met at the university, these three are the only ones I know that grew up in the countryside.

37. Andreas (2004, 18). It must be remembered that the number of years of education that a senior middle school degree entailed had dropped from twelve to ten (or nine in many rural areas).

38. Interviewee 65.

39. Interviewee 82.

40. Interviewee 50.

41. Interviewee 49.

42. Andreas (2004).

CHAPTER 9

1. Zhang, Yao, and Wang were arrested after being summoned to a meeting of the Standing Committee of the CCP Political Bureau, of which they were members. In his memoir, Wu De recounted how the arrests were planned by a small group of party and military leaders (Wu 2004). Also see Forster (1992a); Onate (1978); and Teiwes and Sun (2007).

2. The arrest of the radicals was followed by contention between factions led by Hua Guofeng, who wanted to retain many Mao-era policies, and Deng Xiaoping, who favored a decisive break with the past. By the end of 1978, Deng's faction had prevailed. See Baum (1994); and MacFarquhar (1991).

3. *QHZB* (May 20, 1977); and *XQH* (Oct. 14, 1980).

4. Jiang (1998, 900–904). The Eleventh CCP Congress was held in August 1977.

5. Deng (1984, 241–42).

6. MacFarquhar (1991, 385); and Meisner (1996, 112).

7. Deng quoted in Meisner (1996, 181).

8. Han (2000, 165).

9. Interviewee 55.

10. Ci (1994).

11. Liu (1990, xvii, 22).

12. Interviewee 33.

13. Wan (1987, 99–100).

14. Liu et al. (1987, 162–64).

15. Deng quoted in Teng and Huang (2003).

16. Dickson (2003, 31).

17. Li (2001); and Burns (1989).

18. Ch'i (1991) and Rosen (1992) discuss changing attitudes about party membership.

19. Deng (1984, 53).

20. Liu and Fang (1998, 648–50).

21. Interviewee 22.

22. A semiofficial history of Tsinghua and Peking universities suggested that most of the new students had attended middle school before the Cultural Revolution (Liu and Fang 1998, 660). Cai Jianshe, a graduate of Tsinghua's attached middle school who tested into Tsinghua University in 1977, provided a detailed account of the family backgrounds of his thirty-six classmates. Half were from intellectual families and another 17 percent were children of government officials (Interviewee 22).

23. Interviewee 29.

24. Liu and Fang (1998, 654–55).

25. Liu and Fang (1998, 657–751).

26. State Education Commission (1984, 196–97).

27. Andreas (2004, 18).

28. For analyses of the contraction of middle school education and the renewed emphasis on keypoint schools, see Bakken (1988); Han (2000); Pepper (1996); Rosen (1987); and Thogersen (1990). Pepper (1994) put together a collection of documents that explain the official rationale of the contraction.

29. Kipnis (2001), Anagnost (2004), and Kipnis (2006) analyzed the campaigns to improve the quality of education as part of wider efforts to improve the *suzhi* of the Chinese population, in which *suzhi* is identified with the qualities of the urban middle class.

30. This figure is based on the number of incoming students who changed their household registration from rural to urban between 1995 and 1999, the most recent years for which data was available. The figures were recorded in the *Beijing shi gaoxiao luqu xinsheng banli huji qingkuang tongji biao* (City of Beijing college new student enrollment household registration situation statistics form), which Tsinghua submits to Beijing authorities.

31. Liu et al. (1987, 156).

32. Interviewee 77.

33. *XQH* (Oct. 14, 1980, 2).

34. *XQH* (Oct. 14, 1980, 6).

35. *QHXYTX* (No. 12, Oct. 1985, 147).

36. Zhao (2000, 2).

37. For information about Tsinghua's School of Economics and Management, see http://www.sem.tsinghua.edu.cn. See also Chandler (2005).

38. "Wal-Mart Donating US$1 Million to Establish Tsinghua University China Retail Research Center," Nov. 2, 2004, http://www.wal-martchina.com/english/news/20041102.htm.

39. Eleventh Central Committee (1981, 37).
40. Hu (1981).
41. Jiang (1998, 957).
42. *XQH* (Oct. 14, 1980, 2).

CHAPTER 10

1. Deng (1993, vol. 2, 262).
2. Lee (1991, 302–8).
3. Dickson (2003, 34).
4. Walder (2004, 198). By conducting a life-history survey, Walder and his colleagues were able to compare recruitment rates during different periods.
5. Interviewee 66.
6. Rofel (1999).
7. Hu quoted in Lee (1991, 231).
8. Dreyer (1996); Kamphausen (2007); and Li (2001).
9. Li and White (1990, 15–16).
10. Deng (1984, 308).
11. Lee (1991, 306–7).
12. Manion (1993, 78).
13. Li and White (1990, 14).
14. Lee (1991, 256). The increase in the number of college-educated officials was in part accomplished by sending cadres to back to school, but large numbers of cadres were retired and replaced by better-educated cadres.
15. Li and White (2003, 578).
16. Data on public sector workers is from National Bureau of Statistics of China (2006, 128–29). I estimated the number of cadres in 1990 based on data in Heilmann and Kirchberger (2000, 11–13).
17. Heilmann and Kirchberger (2000, 11). Some of the remaining older members of the political elite do not have university degrees.
18. Li and White (1998); Li (2001); and Li (2002, 2–3).
19. Li (2001, 140–41). Li's father and Jiang's adoptive father were killed.
20. McCarthy (1999).
21. The career trajectories of Hu, Wen, and Wu can be found at www.chinavitae.com. See also Li (2001); Wan (2006); Tkacik (2004); and Yan (2005).
22. Li (2000, 22); and Li and Cheng (2003, 562).
23. The career trajectories of Li and Xi can be found at www.chinavitae.com. See also Li (2007).
24. Li (2007) makes this argument.
25. This motto was inscribed in the CCP's Constitution, at Hu Jintao's behest, in 2007.

CHAPTER 11

1. Information about Tongfang can be found on the company's Web site, www.thtf.com.cn.

2. For biographical information about Lu Zhicheng, see www.thtf.com.cn. For biographical information about Rong Yonglin, see http://www.cec-ceda.org.cn/huodong/2006zscq/ryl.htm.

3. Chen (2006).

4. For Tongfang executive compensation, see http://www.mediapoint.co.uk/investing/quotes. Lu quoted in "Stock Option Creates a Flurry of Chinese Millionaires," http://www.peopledaily.com.cn/english (July 21, 2000).

5. Information about Tsinghua Science Park can be found at http://www.thsp.com.cn.

6. High-technology incubators operated by universities are discussed in Eun, Lee, and Wu (2006); Harwit (2002); and Sunami (2002).

7. Naughton (2007); and Garnaut, Song, and Yao (2006).

8. Li and Rozelle (2003); and Garnaut, Song, and Yao (2006).

9. National Bureau of Statistics of China (2006), 128–29. The public sector includes both state-owned and collective enterprises.

10. Naughton (2007).

11. Solinger (2003, 69).

12. Solinger (1995); and Wang (2006a).

13. Among sixty-three successful private entrepreneurs in Shanxi Province interviewed by Goodman, for instance, twenty-five were children of a party member, most of whom were "in positions of some responsibility within the party-state" (Goodman 2004, 159).

14. Balfour, Einhorn, and Murphy (2003); and Lam (2004).

15. Oster (2006).

16. *China Vitae* (http://www.chinavitae.org); and Chen (2008).

17. *China Vitae* (http://www.chinavitae.org); and Tkacik (2004).

18. Reuters (2003).

19. For biographical sketches of several successful entrepreneurs, see Goodman (2004).

20. The Hurun Report estimated that only one-third of the eight hundred multimillionaires on its 2007 list were party members (http://www.hurun.net). Searching the Hurun Report and three other popular lists of China's wealthiest individuals over the past five years, China's University Alumni Association found only three hundred university graduates, and—unlike top state officials—most of them had not attended the country's most elite schools (Xinhua 2008). Party membership and university degrees are far more common among this group of wealthy entrepreneurs than among the general population, but far less common than among upper-level state officials.

21. Chang and Wong (2004); Naughton (2004); and Yusuf, Nabeshima, and Perkins (2006, 41).

22. The *Almanac of Private Property in China 2000* reported that 19.8 percent of the owners of Chinese private enterprises were party members (Holbig 2002). In a 2002–3 national survey conducted by Tsai (2007), more than one-third of business owners reported they were party members. Among the owners of large firms in four provinces surveyed by Dickson (2007), half were party members and most of the rest wanted to join; the larger the enterprise, the more likely the owner was to be a member. Among sixty-three successful private entrepreneurs in Shanxi Province interviewed by Goodman (2004), 39 percent were party members.

23. Goodman (2003); and Dickson (2007).

24. Zhang, Zhao, Park, and Song (2005, 739).

25. Goodman and Zang (2008). See also Naughton (2007, 217–21); and Wang (2006b).

26. Fang quoted in Kraus (1989, 299).

27. Qian quoted in Cheng and White (1991, 362).

28. Buckley (1991), Hamrin (1990), Kraus (1989), and Li and White (1991) provided insightful accounts of the popularity of meritocratic and technocratic ideas in the 1980s.

29. Among the best-known technocratic scenarios published in the West during this period were those by Bell (1973), Galbraith (1967), and Gouldner (1979). Bell concurred that the rise of planning meant the ascent of the technical and professional intelligentsia, but he argued that social selection would never be based entirely on knowledge. Planning, he wrote, required adjudicating among competing interests and values, a process that ultimately was political, not technical, and which, therefore, would empower politicians, whose principal source of power was not technical competence, but rather the machinery of a political party. "It is not the technocrat who ultimately holds power," wrote Bell, "but the politician" (Bell 1973, 360).

30. Wei (2007).

CONCLUSION

1. Bailes (1978).

2. Lenin quoted in Lampert (1979, 12). See also Lewin (1985, 231); Fitzpatrick (1979, 87); Bailes (1978, 19–23); and Karabel (1997).

3. For the number of engineers in the party, see Bailes (1978, 138). Rigby (1968) provided data on the number of party members (p. 52) and the proportion of worker origin (p. 116). Many working-class party members had been promoted to managerial positions.

4. Bailes (1978, 64–66); and Lampert (1979, 22–24).

5. Lapidus (1978, 82).

6. All-Russian Communist Party (1953, 111–12).
7. Fitzpatrick (1979); Zhu (1997); and Zhu (2000).
8. Fitzpatrick used this term to identify the period of cultural radicalism between 1928 and 1931, in part, to highlight the similarity of Soviet policies during this period and those of the Chinese Cultural Revolution. The term was widely used in the Soviet Union at the time, but not specifically to identify this period (Fitzpatrick 1999).
9. Bailes (1978, 159–261); Fitzpatrick (1979, 113–211); Lapidus (1978); and Zhu (2000).
10. Timasheff (1946).
11. Bailes (1978, 153–54); and Fitzpatrick (1979, 213–17).
12. Fitzpatrick (1979); and Timasheff (1946).
13. Lewin (1985, 254–55).
14. Fitzpatrick (1992, 149–82); Bailes (1978, 268–87); and Hoffman (1993).
15. Fitzpatrick (1992, 149–82); and Bailes (1978, 267).
16. Bailes (1978, 196).
17. Bailes (1978, 225–26) and Fitzpatrick (1993) show that old elites were specific targets between 1928 and 1930, but not during later purges, including those of the 1936–38 period.
18. Stalin (1976).
19. Fitzpatrick (1992, 15).
20. Bailes (1978, 410–11); and Moore (1965, 241–43).
21. Chinese education policies were influenced by remnants of radical practices that survived in the Soviet education system, and the CCP investigated the evolution of education policy in the Soviet Union (see, for instance, *JGS*, Dec. 7, 1967).
22. Schurmann (1968, 220–308).
23. Kelly and Klein (1981).
24. Andrle (1988); Chase (1987); Fitzpatrick (1994); Getty (1985); Hoffman (1993); Koenker (2005); and Lupher (1996).
25. Timasheff (1946); and Lowenthal (1970). Lowenthal accurately predicted the demise of class leveling in China, based on the Soviet and Yugoslav experiences.

Bibliography

Ahn, Byung-joon. 1974. "The Cultural Revolution and China's Search for Political Order." *China Quarterly* 58:249–85.

All-Russian Communist Party (Bolsheviks). 1953 (1919). "Program of the All-Russian Communist Party (Bolsheviks)." In *Materials for the Study of the Soviet System: State and Party Constitutions, Laws, Decrees, Decisions, and Official Statements of the Leaders in Transition*, edited by James Meisel and Edward Kozera. Ann Arbor, MI: George Wahr Publishing Company.

Anagnost, Ann. 2004. "The Corporeal Politics of Quality (Suzhi)." *Public Culture* 16 (2): 189–208.

Andreas, Joel. 2002. "Battling over Political and Cultural Power During the Chinese Cultural Revolution." *Theory and Society* 31:463–519.

———. 2004. "Leveling the 'Little Pagoda': The Impact of College Examinations—and Their Elimination—on Rural Education in China." *Comparative Education Review* 48 (1): 1–47.

———. 2007. "The Structure of Charismatic Mobilization: A Case Study of Rebellion During the Chinese Cultural Revolution." *American Sociological Review* 72:434–58.

Andrle, Vladimir. 1988. *Workers in Stalin's Russia: Industrialization and Social Change in a Planned Economy*. New York: St. Martin's Press.

Bailes, Kendall. 1978. *Technology and Society Under Lenin and Stalin: Origins of the Soviet Technical Intelligentsia, 1917–1941*. Princeton, NJ: Princeton University Press.

Bakken, Borge. 1988. "Backward Reform in Chinese Education." *Australian Journal of Chinese Affairs* 19/20:127–63.

Balfour, Frederik, Bruce Einhorn, and Kate Murphy. 2003. "A Bush in Hand Is Worth . . . a Lot." *Business Week* (Dec. 15): 56.

Barnett, A. Doak. 1967. *Cadres, Bureaucracy, and Political Power in Communist China*. New York: Columbia University Press.

Barnouin, Barbara, and Changgen Yu. 1993. *Ten Years of Turbulence: The Chinese Cultural Revolution*. London: Kegan Paul International.

Baum, Richard. 1994. *Burying Mao: Chinese Politics in the Age of Deng Xiaoping*. Princeton, NJ: Princeton University Press.

Baylis, Thomas. 1974. *The Technical Intelligentsia and the East German Elite: Legitimacy and Social Change in Mature Communism*. Berkeley: University of California Press.

Becker, Gary. 1964. *Human Capital: A Theoretical and Empirical Analysis, with Special Reference to Education*. Chicago: University of Chicago Press.

Beijing Forestry Institute Red Guard Fighting Groups. 1966. *Yi pian hen hao de fanmian jiaocai* (A good piece of negative teaching material). Beijing: Beijing Forestry Institute "East Is Red Commune."

Bell, Daniel. 1973. *The Coming of Post-Industrial Society*. New York: Basic Books.

Bian, Yanjie, Xiaoling Shu, and John Logan. 2001. "Communist Party Membership and Regime Dynamics in China." *Social Forces* 79 (3): 805–41.

Big Criticism Group of Beijing University and Tsinghua University. 1975. "Jiaoyu geming bu rongyi cuangai" (It is not easy to divert the education revolution). *Hongqi* 12.

Bourdieu, Pierre. 1984. *Distinction*. Cambridge, MA: Harvard University Press.

———. 1986. "The Forms of Capital." In *Handbook of Theory and Research for the Sociology of Education*, edited by John Richardson, 241–58. New York: Greenwood Publishing Group.

———. 1989. *The State Nobility*. Stanford, CA: Stanford University Press.

———. 1998. "The 'Soviet' Variant of Political Capital." *Practical Reason: On the Theory of Action*, 14–18. Stanford, CA: Stanford University Press.

Bratton, Dale. 1979. "University Admissions Policies in China, 1970–1978." *Asian Survey* 19 (10): 1008–22.

Broaded, C. Montgomery. 1990. "The Lost and Found Generation: Cohort Succession in Chinese Higher Education." *Australian Journal of Chinese Affairs* 23:77–95.

Brugger, William. 1976. *Democracy and Organization in the Chinese Industrial Enterprise, 1949–1953*. Cambridge: Cambridge University Press.

Bu Weihua. 1998. "Qinghua Fuzhong Hongweibing chengli shimo" (The whole story of the founding of the Tsinghua Attached Middle School Red Guards). *Zhonggong dangshi ziliao* (Materials on the history of the Chinese Communist Party), 70:96–107. Beijing: Communist Party History Press.

Buckley, Christopher. 1991. "Science as Politics and Politics as Science: Fang Lizhi and Chinese Intellectuals' Uncertain Road to Dissent." *Australian Journal of Chinese Affairs* 25:1–36.

Burawoy, Michael. 1998. "The Extended Case Method." *Sociological Theory* 6:14–33.

Burns, John, ed. 1989. *The Chinese Communist Party's* Nomenklatura *System: A Documentary Study of Party Control of Leadership Selection, 1979–1984*. Armonk, NY: M. E. Sharpe.

Carlisle, Robert. 1987. *The Proffered Crown: Saint-Simonianism and the Doctrine of Hope*. Baltimore: Johns Hopkins University Press.

Chamberlain, Heath. 1972. "Transition and Consolidation in Urban China: A Study of Leaders and Organizations in Three Cities, 1949–53." In *Elites in the People's Republic of China*, edited by Robert Scalapino, 245–301. Seattle: University of Washington Press.

Chan, Anita. 1985. *Children of Mao: Personality Developments and Political Activism in the Red Guard Generation*. London: Macmillan.

Chan, Anita, Richard Madsen, and Jonathan Unger. 1984. *Chen Village: The Recent History of a Peasant Community in Mao's China*. Berkeley: University of California Press.

Chan, Anita, Stanley Rosen, and Jonathan Unger. 1980. "Students and Class Warfare: The Roots of the Red Guard Conflict in Guangzhou." *China Quarterly* 3:397–446.

———, eds. 1985. *On Socialist Democracy and the Chinese Legal System: The Li Yizhe Debates*. Armonk, NY: M. E. Sharpe.

Chandler, Clay. 2005. "From Marx to Market: How China's Best Business School Is Rewiring the Nation's Economy." *Fortune* (May 16): 102–12.

Chang, Eric, and Sonia Wong. 2004. "Political Control and Performance in China's Listed Firms." *Journal of Comparative Economics* 32 (4): 617–36.

Chang, Jung, and Jon Halliday. 2005. *Mao: The Unknown Story*. New York: Alfred A. Knopf.

Chang, Parris. 1979. "Who Gets What, When and How in Chinese Politics: A Case Study of the Strategies of Conflict of the 'Gang of Four.'" *Australian Journal of Chinese Affairs* 2:21–42.

Chase, William. 1987. *Workers, Society, and the Soviet State: Labor and Life in Moscow, 1918–1929*. Urbana: University of Illinois Press.

Chen, George. 2008. "Morgan Stanley May Gain at End of China Venture." *International Herald Tribune* (Jan. 15).

Chen, Shu-Ching Jean. 2006. "China's First Son Keeps Low, Goes Global." Forbes .com (Dec. 13). http://www.forbes.com/facesinthenews/2006/12/13/hu-haifeng -china-markets-equity-cx_jc_1213autofacescan02.html.

Chen, Theodore. 1960. *Thought Reform of Chinese Intellectuals*. Hong Kong: Hong Kong University Press.

———. 1981. *Chinese Education Since 1949: Academic and Revolutionary Models*. New York: Pergamon.

Ch'i, His-Sheng. 1991. *Politics of Disillusionment: The Chinese Communist Party Under Deng Xiaoping, 1978–1989*. Armonk, NY: M. E. Sharpe.

Chi Qun, and Xie Jingyi. 1976. *Chi Qun, Xie Jingyi zai Qinghua Daxue Jixie Xi xueyuan he ganbu xuexi 1976 nian 5 yue 16 ri liang bao yi kan shelun zuotanhui shang de jianghua (yuan jian)* (Chi Qun and Xie Jingyi speak at a meeting to study commentaries published on May 16, 1976, by the two-newspapers-and-one-journal with students and cadres of Tsinghua University's Mechanical Engineering Department [primary document]). Beijing.

China Youth Press. 1958. *Lun youhong youzhuan* (On Red and expert). Beijing: Chinese Youth Press.

Chinese Academy of Social Sciences Population Research Center. 1985. *Zhongguo renkou nianjian* (Chinese population yearbook). Beijing: Chinese Social Science Press.

Ci, Jiwei. 1994. *Dialectic of the Chinese Revolution: From Utopianism to Hedonism*. Stanford, CA: Stanford University Press.

Cleverly, John. 1985. *The Schooling of China*. London: George Allen and Unwin.

———. 2000. *In the Lap of Tigers: The Communist Labor University of Jiangxi Province*. Lanham, MD: Rowman and Littlefield.

Cong, Xiaoping. 2007. *Teachers' Schools and the Making of the Modern Chinese Nation-State, 1897–1937*. Vancouver: University of British Columbia Press.

Cui Xianglu, ed. 1993. *Dongfang jiaoyu de jueqi: Mao Zedong jiaoyu sixiang yu Zhongguo jiaoyu 70 nian* (The rise of oriental education: Mao Zedong's educational thinking and 70 years of education in China). Zhengzhou: Henan Education Press.

Deng Xiaoping. 1984. *Selected Works of Deng Xiaoping (1975–1982)*. Beijing: Foreign Languages Press.

———. 1993. *Deng Xiaoping wenxuan* (Selected works of Deng Xiaoping). Beijing: Peoples Press.

Deng, Zhong, and Don Treiman. 1997. "The Impact of the Cultural Revolution on Trends in Educational Attainment in the People's Republic of China." *American Journal of Sociology* 103 (2): 391–428.

Dickson, Bruce. 2003. *Red Capitalists: The Party, Private Entrepreneurs, and Prospects for Political Change*. Cambridge: Cambridge University Press.

———. 2007. "Integrating Wealth and Power in China: The Communist Party's Embrace of the Private Sector." *China Quarterly* 192:827–54.

Dickson, Bruce, and Maria Rost Rublee. 2000. "Membership Has Its Privileges: The Socioeconomic Characteristics of Communist Party Members in Urban China." *Comparative Political Studies* 33 (1): 87–112.

Ding Shu. 2000. "Jinru Zhonggong zhongyang hexin de yidianyuan" (Telegram decoder who entered the center of power in the CCP). *Huaxia wenzhai, Wenge bowuguan tongxun* (China news digest, Cultural Revolution archives correspondence) 65. http://www.cnd.org/cr/ZK00/zk210.hz8.html.

Ding Wei. 2007. "TEEC: Qinghua qiyejia quan" (TEEC: The Tsinghua circle of entrepreneurs). *Zhongguo qiyejia* (China entrepreneur) 7. http://www.cnki.com.cn/Article/CJFDTotal-ZGQY200707027.htm.

Dittmer, Lowell. 1978. "Bases of Power in Chinese Politics: A Theory and an Analysis of the Fall of the 'Gang of Four.'" *World Politics* 31 (1): 26–60.

———. 1987. *China's Continuous Revolution: The Post-Liberation Epoch, 1949–1981.* Berkeley: University of California Press.

———. 1998. *Liu Shaoqi and the Chinese Cultural Revolution: The Politics of Mass Criticism.* Berkeley: University of California Press.

Djilas, Milovan. 1957. *The New Class: An Analysis of the Communist System.* New York: Praeger.

Dreyer, June. 1996. "The New Officer Corps: Implications for the Future." *China Quarterly* 146:315–35.

Eleventh Central Committee of the Communist Party of China. 1981. "On Questions of Party History: Resolution on Certain Questions in the History of Our Party Since the Founding of the People's Republic of China." *Beijing Review* 27:10–39.

Endicott, Stephen. 1988. *Red Earth: Revolution in a Sichuan Village.* London: I. B. Taurus.

Eun, Jong-Hak, Keun Lee, and Guisheng Wu. 2006. "Explaining the 'University-run enterprises' in China: A Theoretical Framework for University-Industry Relationship in Developing Countries and Its Application to China." *Research Policy* 35 (9): 1329–46.

Eyal, Gil, Ivan Szelenyi, and Eleanor Townsley. 1998. *Making Capitalism Without Capitalists: The New Ruling Elites in Eastern Europe.* London: Verso Books.

Fang Huijian and Zhang Sijing, eds. 2001. *Qinghua Daxue zhi* (Annals of Tsinghua University), 2 vols. Beijing: Tsinghua University Press.

Fang Huijian and Hao Weiqian, eds. 1999. *Jiang Nanxiang jiaoyu sixiang yanjiu* (Research on Jiang Nanxiang's education thinking). Beijing: Tsinghua University Press.

Fitzpatrick, Sheila. 1979. *Education and Social Mobility in the Soviet Union, 1921–1934.* New York: Cambridge University Press.

———. 1992. *The Cultural Front: Power and Culture in Revolutionary Russia.* Ithaca, NY: Cornell University Press.

———. 1993. "The Impact of the Great Purges on Soviet Elites: A Case Study from Moscow and Leningrad Telephone Directories of the 1930s." In *Stalinist Terror: New Perspectives,* edited by J. Arch Getty and Roberta Manning, 247–60. Cambridge: Cambridge University Press.

———. 1994. "Workers Against Bosses: The Impact of the Great Purges on Labor-Management Relations." In *Making Workers Soviet: Power, Class and Identity,* edited by Lewis Siegelbaum and Ronald Grigor Suny, 311–40. Ithaca, NY: Cornell University Press.

———. 1999. "Cultural Revolution Revisited." *Russian Review* 58 (2): 202–9.

Foreign Languages Department Revolutionary Committee. 1968. *"Wuqi" Hanying ciyu huipian* (The "May 7" Chinese-English collection of terms and expressions). Wuhan: Huazhong Teachers College Revolutionary Committee.

Forster, Keith. 1990. *Rebellion and Factionalism in a Chinese Province: Zhejiang, 1966–1976*. Armonk, NY: M. E. Sharpe.

———. 1992a. "China's Coup of October 1976." *Modern China* 18 (3): 263–303.

———. 1992b. "Spontaneous and Institutional Rebellion in the Cultural Revolution: The Extraordinary Case of Weng Senhe." *Australian Journal of Chinese Affairs* 27:39–75.

Friedman, Edward, Paul Pickowicz, and Mark Selden. 1991. *Chinese Village, Socialist State*. New Haven, CT: Yale University Press.

Funnell, Victor. 1970. "The Chinese Communist Youth Movement, 1949–1966." *China Quarterly* 42:105–30.

Galbraith, John K. 1967. *The New Industrial State*. Boston: Houghton Mifflin.

Gao, Mobo. 1999. *Gao Village: Rural Life in Modern China*. London: Hurst and Co.

Gardner, John. 1969. "The *Wu-fan* Campaign in Shanghai: A Study in the Consolidation of Urban Control." In *Chinese Communist Politics in Action*, edited by A. Doak Barnett, 477–539. Seattle: University of Washington Press.

Gaurnaut, Ross, Ligang Song, and Yang Yao. 2006. "Impact and Significance of State-Owned Enterprise Restructuring in China." *The China Journal* 55:35–63.

Getty, J. Arch. 1985. *The Origins of the Great Purges: The Soviet Communist Party Reconsidered, 1933–1938*. Cambridge: Cambridge University Press.

Goldman, Merle. 1967. *Literary Dissent in Communist China*. Cambridge, MA: Harvard University Press.

———. 1970. "Party Policies Towards Intellectuals: The Unique Blooming and Contending of 1961–2." In *Party Leadership and Revolutionary Power in China*, edited by John Lewis, 268–303. Cambridge: Cambridge University Press.

———. 1981. *China's Intellectuals: Advice and Consent*. Cambridge, MA: Harvard University Press.

Goldman, Merle, Timothy Cheek, and Carol Lee Hamrin. 1987. *China's Intellectuals and the State: In Search of a New Relationship*. Cambridge, MA: Harvard University Council on East Asian Studies.

Goldstein, Avery. 1991. *From Bandwagon to Balance-of-Power Politics: Structural Constraints and Politics in China, 1949–1978*. Stanford, CA: Stanford University Press.

Goodman, David. 2003. "New Entrepreneurs in Reform China: Economic Growth and Social Change in Taiyuan, Shanxi." In *Capital and Knowledge in Asia: Changing Power Relations*, edited by Heidi Dahles and Otto van den Muijzenberg, 187–97. London: Routledge Curzon.

———. 2004. "Localism and Entrepreneurship: History, Identity and Solidarity as Factors of Production." In *China's Rational Entrepreneurs: The Development of the New Private Business Sector*, edited by Barbara Krug, 139–65. London: Routledge Curzon.

Goodman, David, and Xiaowei Zang. 2008. "The New Rich in China: The Dimensions of Social Change." In *The New Rich in China: Future Rulers, Present Lives*, edited by David Goodman, 1–21. London: Routledge.

Gouldner, Alvin. 1979. *The Future of Intellectuals and the Rise of the New Class*. New York: Seabury Press.

Gu, Edward, and Merle Goldman. 2004. *Chinese Intellectuals Between the State and Market*. London: Routledge Curzon.

Gu, Mingyuan. 1984. "The Development and Reform of Higher Education in China." *Comparative Education Review* 20 (1): 141–48.

Hamrin, Carol. 1990. *China and the Challenge of the Future: Changing Political Patterns*. Boulder, CO: Westview Press.

Hamrin, Carol, and Timothy Cheek, eds. 1986. *China's Establishment Intellectuals*. Armonk, NY: M. E. Sharpe.

Han, Dongping. 2000. *The Unknown Cultural Revolution*. New York: Garland Publishing.

Hannum, Emily. 1999. "Political Change and the Urban-Rural Gap in Basic Education in China, 1949–1990." *Comparative Education Review* 43 (2): 193–211.

Hannum, Emily, and Yu Xie. 1994. "Trends in Educational Gender Inequality in China: 1949–1985." *Research in Social Stratification and Mobility* 13:73–98.

Harding, Harry. 1981. *Organizing China: The Problem of Bureaucracy, 1949–1976*. Stanford, CA: Stanford University Press.

———. 1991. "The Chinese State in Crisis." In *The People's Republic, Part 2: Revolutions Within the Chinese Revolution, 1966–82*, vol. 15 of *The Cambridge History of China*, edited by Roderick MacFarquhar and John Fairbank, 107–217. Cambridge: Cambridge University Press.

Harper, Paul. 1971. "Trade Union Cultivation of Workers for Leadership." In *The City in Communist China*, edited by John Lewis, 123–52. Stanford, CA: Stanford University Press.

Harwit, Eric. 2002. "High Technology Incubators: Fuel for China's New Entrepreneurship?" *China Business Review* 29 (4): 26–29.

Hayhoe, Ruth. 1996. *China's Universities 1895–1995: A Century of Cultural Conflict*. New York: Garland Publishing.

He Chongling, ed. 2001. *Qinghua Daxue jiushi nian* (Tsinghua University ninety years). Beijing: Tsinghua University Press.

Heartfield, James. 2005. "China's Comprador Capitalism Is Coming Home." *Review of Radical Political Economics* 37 (2): 196–214.

Heilmann, Sebastian, and Sarah Kirchberger. 2000. *The Chinese* Nomenklatura *in Transition: A Study Based on Internal Cadre Statistics of the Central Organization Department of the Chinese Communist Party*. Trier, Germany: Center for East Asian and Pacific Studies, Trier University.

Hinton, William. 1966. *Fanshen: A Documentary of Revolution in a Chinese Village*. New York: Vintage Books.

———. 1972. *Hundred Day War: The Cultural Revolution at Tsinghua University.* New York: Monthly Review Press.

Hoffman, David. 1993. "The Great Terror on the Local Level: Purges in Moscow Factories, 1936–1938." In *Stalinist Terror: New Perspectives*, edited by J. Arch Getty and Roberta Manning, 116–65. Cambridge: Cambridge University Press.

Holbig, Heike. 2002. "The Party and Private Entrepreneurs in the PRC." *Copenhagen Journal of Asian Studies* 16:30–56.

Hu Ping. 1981. "On the Question of Intellectuals." *Beijing Review* 7:13–17.

Huang, Jing. 2000. *Factionalism in Chinese Communist Politics.* Cambridge: Cambridge University Press.

Inkeles, Alex. 1966. "Social Stratification and Mobility in the Soviet Union." In *Class, Status and Power: Social Stratification in Comparative Perspective*, edited by Reinhard Bendix and Seymour Lipset. New York: Free Press.

Israel, John. 1982–83. "Introduction to the Draft History of Tsinghua University." *Chinese Education* 15 (3/4): iv–xv.

Israel, John, and Donald Klein. 1976. *Rebels and Bureaucrats: China's December 9ers.* Berkeley: University of California Press.

Jiang Nanxiang. 1967. *Jiaodai cailiao (gong pipan)* (Confession materials [for criticism]). Beijing: Manuscript distributed by Tsinghua University Jinggangshan—April 14.

———. 1998. *Jiang Nanxiang wenji* (Collected works of Jiang Nanxiang). Beijing: Tsinghua University Press.

Jiang Qing, Yao Wenyuan, Chi Qun, and Xie Jingyi. 1976. *Jiang Qing, Yao Wenyuan, Chi Qun, Xie Jingyi 1974 nian 1 yue 25 ri zai zhongyang zhishu jiguan he guojia jiguan pilin pikong dongyuan dahui shang de jianghua (jilu gao)* (Jiang Qing, Yao Wenyuan, Chi Qun, and Xie Jingyi speak at the January 25, 1974, mass mobilizing meeting convened by central and national agencies to criticize Lin Biao and Confucius [minutes]). Beijing.

Jiang Yiwei. 1957. "Jishu yu zhengzhi" (Technology and politics). *Xuexi* (Study) 16:11–14.

Joint Publications Research Service, ed. 1974. "Dialogues of Responsible Persons of Capital Red Guards Congress (28 July 1968)." In *Miscellany of Mao Tse-tung Thought (1949–1968)*, Part 1, 469–97. Springfield, VA: National Technical Information Service.

Kamphausen, Roy. 2007. "ROTC with Chinese Characteristics: Training the PLA in Civilian Universities." *China Brief* 7 (6): 2–5.

Karabel, Jerome. 1997. "Lenin and the Problem of the Intelligentsia." *Current Perspectives in Social Theory* 17:261–312.

Kau, Ying-Mao. 1969. "The Urban Bureaucratic Elite in Communist China: A Case Study of Wuhan, 1949–1965." In *Chinese Communist Politics in Action*, edited by A. Doak Barnett, 216–77. Seattle: University of Washington Press.

Kelley, Jonathan, and Herbert Klein. 1981. *Revolution and the Rebirth of Inequality: A Theory Applied to the National Revolution in Bolivia*. Berkeley: University of California Press.

Kipnis, Andrew. 2001. "The Disturbing Educational Discipline of Peasants." *The China Journal* 46:1–24.

———. 2006. "*Suzhi*: A Keyword Approach." *China Quarterly* 186: 295–313.

Knight, John, and Shi Li. 1996. "Educational Attainment and the Rural-Urban Divide in China." *Oxford Bulletin of Economics and Statistics* 58 (1): 83–117.

Koenker, Diane. 2005. *Republic of Labor: Russian Printers and Soviet Socialism, 1918–1930*. Ithaca, NY: Cornell University Press.

Konrad, George, and Ivan Szelenyi. 1979. *The Intellectuals on the Road to Class Power*. New York: Harcourt Brace Jovanovich.

———. 1991. "Intellectuals and Domination in Post-Communist Societies." In *Social Theory for a Changing Society*, edited by Pierre Bourdieu and James Coleman, 337–72. Boulder, CO: Westview Press.

Kraus, Richard. 1981. *Class Conflict in Chinese Socialism*. New York: Columbia University Press.

———. 1989. "The Lament of Astrophysicist Fang Lizhi: China's Intellectuals in a Global Context." In *Marxism and the Chinese Experience: Issues in Contemporary Chinese Socialism*, edited by Arif Dirlik and Maurice Meisner, 294–315. Armonk, NY: M. E. Sharpe.

Kuai Dafu. 1966. *Qinghua Daxue dazibao* (Tsinghua University big character posters). Beijing: Tsinghua University Jinggangshan Red Guard Propaganda Team.

Lam, Willy. 2004. "Factional Politics in the CCP." *China Brief* 4 (5): 5–7.

Lampert, Nicholas. 1979. *The Technical Intelligentsia and the Soviet State*. New York: Holmes and Meier.

Lane, David. 1982. *The End of Social Inequality?: Class, Status, and Power Under State Socialism*. London: Allen and Unwin.

Lapidus, Gail Warshofsky. 1978. "Educational Strategies and the Cultural Revolution: The Politics of Soviet Development." In *Cultural Revolution in Russia, 1928–1931*, edited by Sheila Fitzpatrick, 78–104. Bloomington: Indiana University Press.

Leader, Shelah Gilbert. 1974. "The Communist Youth League and the Cultural Revolution." *Asian Survey* 14 (8): 700–715.

Lee, Hong Yung. 1978. *The Politics of the Chinese Cultural Revolution: A Case Study*. Berkeley: University of California Press.

———. 1991. *From Revolutionary Cadres to Technocrats in Socialist China*. Berkeley: University of California Press.

Leijonhufvud, Goran. 1990. *Going Against the Tide: On Dissent and Big Character Posters in China*. London: Curzon Press.

Lewin, Moshe. 1985. *The Making of the Soviet System: Essays in the Social History of Interwar Russia*. New York: Pantheon Books.

Lewis, John. 1963. *Leadership in Communist China*. Ithaca, NY: Cornell University Press.

Li, Cheng. 1994. "University Networks and the Rise of Tsinghua Graduates in China's Leadership." *Australian Journal of Chinese Affairs* 32:1–30.

———. 2000. "Jiang Zemin's Successors: The Rise of the Fourth Generation of Leaders in the PRC." *China Quarterly* 161:1–40.

———. 2001. *China's Leaders: The New Generation*. Lanham, MD: Rowman and Littlefield.

———. 2002. "Hu's Followers: Provincial Leaders with Backgrounds in the Youth League." *China Leadership Monitor* 3. http://www.hoover.org/publications/clm/issues/2906756.html.

———. 2007. "China: Riding Two Horses at Once." http://www.brookings.edu/articles/2007.

Li, Cheng, and Lynn White. 1988. "The Thirteenth Central Committee of the Chinese Communist Party: From Mobilizers to Managers." *Asian Survey* 28 (4): 371–99.

———. 1990. "Elite Transformation and Modern Change in Mainland China and Taiwan: Empirical Data and the Theory of Technocracy." *China Quarterly* 121:1–35.

———. 1991. "China's Technocratic Movement and the *World Economic Herald*." *Modern China* 17 (3): 342–88.

———. 1998. "The Fifteenth Central Committee of the Chinese Communist Party: Full-Fledged Technocratic Leadership with Partial Control by Jiang Zemin." *Asian Survey* 38 (3): 231–64.

———. 2003. "The Sixteenth Central Committee of the Chinese Communist Party: Hu Gets What?" *Asian Survey* 43 (4): 553–97.

Li Hongru and Jiang Xihua. 1994. "Qinghua Daxue xiaoban chanye fazhan guocheng chuyi" (A modest proposal regarding the development process of Tsinghua University school-run productive enterprises). *Qinghua Daxue jiaoyu yanjiu* (Tsinghua University research on education) 1:82–87.

Lieberthal, Kenneth. 1980. *Revolution and Tradition in Tientsin, 1949–1952*. Stanford, CA: Stanford University Press.

———. 1995. *Governing China: From Revolution Through Reform*. New York: W. W. Norton.

Liu Bing. 1998. *Fengyu suiyue: Qinghua Daxue "Wenhua Da Geming" yishi* (Stormy years: Recollections of the "Cultural Revolution" at Tsinghua University). Beijing: Tsinghua University Press.

Liu Binyan. 1990. *China's Crisis, China's Hope*. Translated by Howard Goldblatt. Cambridge, MA: Harvard University Press.

Liu Guokai. 1986–87. "A Brief Analysis of the Cultural Revolution." *Chinese Sociology and Anthropology* 19 (2): 14–151.

Liu Jingming. 1999. "Jiaoyu zhidu he jiaoyu huode de daiji yinxiang" (The intergenerational influence of the education system and education attainment). In *Shengming de licheng: zhongda shehui shijian yu zhongguo ren de shengming guiji* (Life course: Major social events and Chinese people's life paths), edited by Li Qiang. Hangzhou: Zhejiang People's Press.

Liu Kexuan and Fang Mingdong, eds. 1998. *Beida yu Qinghua* (Peking University and Tsinghua). Beijing: National Administrative Institute Press.

Liu Shao-chi. 1967. *How to Be a Good Communist.* Boulder, CO: Panther Publications.

Liu Shuli et al., eds. 1987. *Qinghua Daxue xiaoshi gangyao (taolun gao)* (Tsinghua University outline school history [discussion draft]). Beijing: School History Editing and Writing Group.

Lowenthal, Richard. 1970. "Development vs. Utopia in Communist Policy." In *Change in Communist Systems,* edited by Chalmers Johnson, 33–116. Stanford, CA: Stanford University Press.

Ludz, Peter. 1972. *The Changing Party Elite in East Germany.* Cambridge, MA: MIT Press.

Lupher, Mark. 1996. *Power Restructuring in China and Russia.* Boulder, CO: Westview Press.

Ma Yanwen. 1976. "Wuchan jieji zhuanzheng yu guanliao zhuyizhe jieceng: Xuexi Mao Zhuxi zhongyao zhishi tihui zhici" (The bureaucratic class and the dictatorship of the proletariat: Understandings gained from studying Chairman Mao's important instructions, part 4). *Beijing Daxue xuebao* (Peking University journal) 4:3–12.

MacFarquhar, Roderick. 1960. *The Hundred Flowers.* London: Stevens and Sons.

———. 1974. *The Origins of the Cultural Revolution. Vol. 1: Contradictions Among the People, 1956–1957.* Oxford: Oxford University Press.

———. 1991. "The Succession to Mao and the End of Maoism." In *The People's Republic, Part 2: Revolutions Within the Chinese Revolution, 1966–1982,* vol. 15 of *The Cambridge History of China,* edited by John Fairbank and Roderick MacFarquhar, 305–401. Cambridge: Cambridge University Press.

MacFarquhar, Roderick, and Michael Schoenhals. 2006. *Mao's Last Revolution.* Cambridge, MA: Harvard University Press.

MacIver, Alice. 1922. "Saint-Simon and His Influence on Marx." *Economica* 6:238–45.

Manion, Melanie. 1993. *Retirement of Revolutionaries: Public Policies, Social Norms, Private Interests.* Princeton, NJ: Princeton University Press.

Mann, Michael. 1986. "The Autonomous Power of States: Its Origins, Mechanisms, and Results." In *States in History,* edited by John A. Hall, 109–36. London: Basil Blackwell.

Manuel, Frank. 1956. *The New World of Henri Saint-Simon.* Cambridge, MA: Harvard University Press.

Mao Zedong. 1965 (1927). "Report on an Investigation of the Peasant Movement in Hunan." *Selected Works of Mao Tsetung*, vol. 1: 23–29. Beijing: Foreign Languages Press.

———. 1966. "A Letter to the Red Guards of the Middle School Attached to Tsinghua University." Translated in *Current Background* 891 (1969): 63.

———. 1974 (1966). "Remarks at the Spring Festival." In *Chairman Mao Talks to the People: Talks and Letters: 1956–1971*, edited by Stuart Schram, 197–211. New York: Pantheon Books.

———. 1977 (1958–60). *A Critique of Soviet Economics.* Translated by Moss Roberts. New York: Monthly Review Press.

———. 1996. *Jianguo yilai Mao Zedong wengao* (Writings of Mao Zedong since the founding of the People's Republic of China). 11 vols. Beijing: Central Documents Press.

Marx, Karl. 1973 [1857]. *Grundrisse: Foundations of the Critique of Political Economy.* New York: Penguin.

———. 1978 [1875]. "Critique of the Gotha Program." In *The Marx-Engels Reader*, edited by Robert Tucker, 2nd ed., 525–41. New York: W. W. Norton.

McCarthy, Terry. 1999. "Red Star." *Time* (April 12): 61–62.

Meisner, Maurice. 1982. *Marxism, Maoism, and Utopianism: Eight Essays.* Madison: University of Wisconsin Press.

———. 1986. *Mao's China and After: A History of the People's Republic.* New York: Free Press.

———. 1996. *The Deng Xiaoping Era: An Inquiry into the Fate of Chinese Socialism, 1978–1994.* New York: Hill and Wang.

Montaperto, Ronald. 1972. "From Revolutionary Successors to Revolutionaries: Chinese Students in the Early Stages of the Cultural Revolution." In *Elites in the People's Republic of China*, edited by Robert Scalapino, 575–605. Seattle: University of Washington Press.

Moore, Barrington. 1965. *Soviet Politics: The Dilemma of Power.* New York: Harper and Row.

Mu, Fu-sheng. 1963. *The Wilting of the Hundred Flowers: The Chinese Intelligentsia Under Mao.* New York: Praeger.

National Bureau of Statistics of China. 2006. *Zhonggua tongji nianjian, 2006* (China statistical yearbook, 2006). Beijing: China Statistics Press.

Naughton, Barry. 2004. "Market Economy, Hierarchy, and Single Party Rule: How Does the Transition Path in China Shape the Emerging Market Economy?" Paper presented at the International Economic Association Meeting in Hong Kong.

———. 2007. *The Chinese Economy: Transitions and Growth.* Cambridge, MA: MIT Press.

New Tsinghua Editorial Committee. 1957. *Hanwei gaodeng jiaoyu he kexue shiye de shehui zhuyi fangxiang: piping Youpai fenzi Qian Weichang de fan gong fan shehui zhuyi yanxing* (Defend the socialist orientation of higher education and scientific institutions: Criticize the antiparty, antisocialist speeches and actions of Rightist Qian Weichang). 2 vols. Beijing: Tsinghua University Press.

North, Robert, with Ithiel de Sola Pool. 1966. "Kuomintang and Chinese Communist Elites." In *World Revolutionary Elites: Studies in Coercive Ideological Movements*, edited by Harold Lasswell and Daniel Lerner, 319–455. Cambridge, MA: MIT Press.

Onate, Andres. 1978. "Hua Kuo-feng and the Arrest of the 'Gang of Four.'" *China Quarterly* 75:540–65.

Oster, Shai. 2006. "Attraction of Business Is Powerful for Ms. Li, China's 'Princelings.'" *Wall Street Journal* (May 15): B1.

Parkin, Frank. 1971. *Class Inequality and Political Order: Social Stratification in Capitalist and Communist Societies.* London: MacGibbon and Kee.

Peking University Education Revolution Group. 1973. "Insist on the Unity of Politics and Vocational Work, Comprehensively Grasp Conditions of Enrollment." *Renmin Ribao* (March 4, 1972). Translated in *Chinese Education* 6 (2): 6–11.

People's Education Press, ed. 1976. *Wuchan jieji jiaoyu geming polang qianjin: Qinghua, Beida jiaoyu geming wenzhang xuanbian* (The proletarian education revolution advances breaking the waves: Selection of articles about the education revolution at Tsinghua and Peking University). Beijing: People's Education Press.

Pepper, Suzanne. 1978. "Education and Revolution: The 'Chinese Model' Revisited." *Asian Survey* 19 (10): 847–90.

———, ed. 1994. "Rural Education (I)." *Chinese Education and Society* 27 (5).

———. 1996. *Radicalism and Education Reform in Twentieth-Century China: The Search for an Ideal Development Model.* Cambridge: Cambridge University Press.

Perry, Elizabeth, and Xun Li. 1997. *Proletarian Power: Shanghai in the Cultural Revolution.* Boulder, CO: Westview Press.

Ren Jianming. 1999. *Qinghua Daxue xuesheng gongzuo lunwenji, di si ji* (Tsinghua University student work reader, fourth collection). Beijing: Tsinghua University Press.

Reuters. 2003. "Daughter of Top Communist Weds Top Capitalist." *Sydney Morning Herald* (Nov. 6).

Rigby, T. H. 1968. *Communist Party Membership in the U.S.S.R., 1917–1967.* Princeton, NJ: Princeton University Press.

Riskin, Carl. 1991. *China's Political Economy: The Quest for Development Since 1949.* New York: Oxford University Press.

Rofel, Lisa. 1999. *Other Modernities: Gendered Yearnings in China After Socialism.* Berkeley: University of California Press.

Rosen, Stanley. 1979. *The Origins and Development of the Red Guard Movement in*

China, 1960–1968. Ph.D. dissertation, Department of Political Science, University of California, Los Angeles.

———. 1982. *Red Guard Factionalism and the Cultural Revolution in Guangzhou (Canton).* Boulder, CO: Westview Press.

———. 1987. "Restoring Key Secondary Schools in Post-Mao China: The Politics of Competition and Educational Quality." In *Policy Implementation in Post-Mao China*, edited by David Lampton, 321–53. Berkeley: University of California Press.

———. 1992. "Students and the State in China: The Crisis in Ideology and Organization." In *State and Society in China: The Consequences of Reform*, edited by Arthur Rosenbaum. Boulder, CO: Westview Press.

Scalapino, Robert. 1972. "The Transition in Chinese Party Leadership: A Comparison of the Eighth and Ninth Central Committees." In *Elites in the People's Republic of China*, edited by Robert Scalapino, 67–148. Seattle: University of Washington Press.

Schapiro, Leonard, and John Lewis. 1969. "The Roles of the Monolithic Party Under the Totalitarian Leader." *China Quarterly* 40:39–64.

Schurmann, Franz. 1968. *Ideology and Organization in Communist China.* Berkeley: University of California Press.

Schwartz, Benjamin. 1968. "The Reign of Virtue: Some Broad Perspectives on Leader and Party in the Cultural Revolution." *China Quarterly* 35:1–17.

Sheehan, Jackie. 1998. *Chinese Workers: A New History*. London: Routledge.

Shen Ruhuai. 2004. *Qinghua Daxue Wenge jishi: Yige Hongweibing de zishu* (Tsinghua University Cultural Revolution chronicle: A Red Guard's memoir). Hong Kong: Epoch Arts Press.

Shirk, Susan. 1982. *Competitive Comrades: Career Incentives and Student Strategies in China.* Berkeley: University of California Press.

Snow, Edgar. 1971. *Red China Today*. New York: Random House.

Solinger, Dorothy. 1995. "The Chinese Work Unit and Transient Labor in the Transition from Socialism." *Modern China* 21 (2): 155–83.

———. 2003. "Chinese Urban Jobs and the WTO." *The China Journal* 49:61–86.

Song, Yongyi. 2004. "The Enduring Legacy of Blood Lineage Theory." *China Rights Forum* 4:13–23. http://hrichina.org/public/PDFs/EnduringLegacy4-2004.pdf.

Song, Yongyi, and Dajin Sun, eds. 1996. *Wenhua Da Geming he ta de yiduan sichao* (Heterodox thinking during the Cultural Revolution). Hong Kong: Countryside Book House.

Stalin, J. V. 1976 (1936). "On the Draft Constitution of the U.S.S.R." *Problems of Leninism*, 795–834. Beijing: Foreign Languages Press.

State Education Commission. 1984. *Zhongguo jiaoyu chengjiu: Tongji ziliao, 1949–1983* (Achievement of education in China: Statistics, 1949–1983). Beijing: People's Education Press.

Statistical Work Dispatch Materials Office. 1957. "1955 nian quanguo zhigong renshu, goucheng yu fenbu de gaikuang" (General situation of employees in 1955: Numbers, composition, and distribution). *Xinhua banyuekan* (New China biweekly) 2:87–89.

Sun Liping. 2002. "Zongtixing ziben yu zhuanxingqi jingying xingcheng" (Comprehensive capital and elite formation in periods of transition). *Zhejiang xuekan* (Zhejiang journal) 3.

Sun Zhongzhi, ed. 1990. *Jiang Nanxiang jinian wenji* (Collection of essays to commemorate Jiang Nanxiang). Beijing: Tsinghua University Press.

Sunami, Atsushi. 2002. "Industry-University Cooperation and University-Affiliated Enterprises in China, a Country Aspiring for Growth on Science and Education—Building New System for Technological Innovation." Tokyo: Research Institute of Economy, Trade and Industry. http://www.rieti.go.jp/en/papers/research-review.

Szelenyi, Ivan. 1986. "The Prospects and Limits of the East European New Class Project: An Auto-critical Reflection on the Intellectuals on the Road to Class Power." *Politics and Society* 15 (2): 103–44.

Szelenyi, Ivan, and Bill Martin, 1988. "The Three Waves of New Class Theories." *Theory and Society* 17:645–67.

Tang Shaojie. 1996. "Qinghua Jinggangshan Bingtuan de xingshuai" (Rise and fall of Tsinghua Jinggangshan Regiment). In *Wenhua Da Geming: Shishi yu yanjiu* (Cultural Revolution: Facts and analysis), edited by Liu Qingfeng, 49–63. Hong Kong: Chinese University Press.

———. 1998. "Cong Qinghua Daxue liang pai kan 'Wenhua Da Geming' zhong qunzhong zuzhi de duili he fenqi" (Antagonisms and differences between 'Cultural Revolution' mass organizations from the perspective of the two factions at Tsinghua University). *Zhonggong Dangshi yanjiu* (Chinese Communist Party history research) 2:69–74.

———. 2003. *Yi ye zhiqiu: Qinghua Daxue 1968 nian 'bai ri da wudou'* (An episode of the Cultural Revolution: The 1968 "hundred day war" at Tsinghua University). Hong Kong: Chinese University Press.

———. 2005. "Qinghua Daxue 'jiaoyu geming' shuping" (Commentary on the "education revolution" at Tsinghua University). *Daxue Renwen* (University humanities) 4.

Taylor, Robert. 1981. *China's Intellectual Dilemma: Politics and University Enrollment, 1949–1978.* Vancouver: University of British Columbia Press.

Teiwes, Frederick. 1976. "The Origins of Rectification: Inner-Party Purges and Education Before Liberation." *China Quarterly* 65:15–53.

———. 1984. *Leadership, Legitimacy, and Conflict in China: From a Charismatic Mao to the Politics of Succession.* Armonk, NY: M. E. Sharpe.

Teiwes, Frederick, and Warren Sun. 2007. *The End of the Maoist Era: Chinese Politics During the Twilight of the Cultural Revolution, 1972–1976.* Armonk, NY: M. E. Sharpe.

Teng Teng and Huang Shenglun. 2003. "Yi Nanxiang tongzhi changdao zhengzhi fudaoyuan zhidu" (Remembering how Comrade Nanxiang initiated the political counselors system). *Tsinghua News* (Dec. 11). http://news.Tsinghua.edu.cn.

Thogersen, Stig. 1990. *Secondary Education in China After Mao: Reform and Social Conflict.* Aarhus, Denmark: Aarhus University Press.

Timasheff, Nicholas. 1946. *The Great Retreat: The Growth and Decline of Communism.* New York: E. P. Dutton.

Tkacik, John. 2004. "Premier Wen and Vice-President Zeng: The 'Two Centers' of China's Fourth Generation." In *Civil-Military Change in China: Elites, Institutes and Ideas After the 16th Party Congress*, edited by Andrew Scobell and Larry Wortzel, 162–63. Washington, DC: Strategic Studies Institute.

Townsend, James. 1970. "Intra-Party Conflict in China: Disintegration in an Established One-Party System." In *Authoritarian Politics in Modern Society*, edited by Samuel Huntington and Clement Moore, 284–310. New York: Basic Books.

Treiman, Donald, et al., eds. 1998. *Life Histories and Social Change in Contemporary China: Provisional Codebook.* Los Angeles: University of California, Los Angeles, Institute for Social Science Research.

Tsai, Kellee. 2007. *Capitalism Without Democracy: The Private Sector in Contemporary China.* Ithaca, NY: Cornell University Press.

Tsinghua Jinggangshan April 14 Revolutionary Link-up Meeting. 1967a. *Dazibao xuanbian* (Selected big character posters). 2 vols. Beijing.

Tsinghua Jinggangshan Regiment. 1967a. *Dazibao xuanbian* (Selected big character posters). Vols. 1, 4, 5, 6, and 7. Beijing.

———. 1967b. "Chushen lun" (On family origin). *Jinggangshan tongxun* (Jinggangshan news report) 138. Beijing.

Tsinghua Jinggangshan United General Headquarters April 14 Cadre Office. 1967. *Dazibao xuanbian* (Selected big character posters), 2 vols. Beijing.

Tsinghua University. 1975. *Gexiang jiben qingkuang tongji* (Statistics on the basic situation of each aspect). Beijing.

Tsinghua University Attached Middle School 80th Anniversary Celebration Committee. 1995. *Jinian wenji* (Collected memoirs). Beijing: Tsinghua University Attached Middle School.

Tsinghua University Attached Middle School Jinggangshan Regiment. 1967. "Tequan jiecheng de weidaoshi" (Apologists for the privileged stratum). *Chun Lei* (Spring Thunder) (July). (Reproduced in *Red Guard Publications.* Washington, D.C.: Center for Chinese Research Materials Association of Research Libraries, 1979, vol. 4: 753.)

Tsinghua University Attached Middle School Red Guards. 1966. "Hail, Line of the Proletarian Class!" *China News Analysis* 63:2–5.

Tsinghua University Attached Middle School Red Guards Senior Middle School Class No. 655 Leadership Small Group. 1996 (1966). "Zuo dingtian lidi de ren"

(Be a dauntless fighter). In *Wenhua Da Geming he ta de yiduan sichao* (Heterodox thinking during the Cultural Revolution), edited by Song Yongyi and Sun Dajin, 86–87. Hong Kong: Countryside Book House.

Tsinghua University President's Office. 1954. *Qinghua Daxue 1954 nian gongzuo baogao* (Tsinghua University 1954 work report). Beijing.

Tsinghua University School History Research Office. 1983, 1995, 1996. *Qinghua renwu zhi* (Eminent persons of Tsinghua University). 4 vols. Beijing: Tsinghua University Press.

Tsinghua University Workers' Peoples Liberation Army Propaganda Team. 1970. "Strive to Build a Socialist University of Science and Engineering." *Peking Review* 31:5–31.

Tsou, Tang. 1969. "The Cultural Revolution and the Chinese Political System." *China Quarterly* 38:63–91.

———. 1995. "Chinese Politics at the Top: Factionalism or Informal Politics? Balance-of-Power Politics or a Game to Win All?" *The China Journal* 34:95–156.

Unger, Jonathan. 1982. *Education Under Mao: Class and Competition in Canton Schools, 1960–1980*. New York: Columbia University Press.

———. 1984. "Severing the Links Between Education and Careers: The Sobering Experience of China's Urban Schools." In *Education Versus Qualifications?*, edited by John Oxenham, 176–91. London: George Allen and Unwin.

Vogel, Ezra. 1967. "From Revolutionary to Semi-Bureaucrat: The 'Regularization' of Cadres." *China Quarterly* 29:36–60.

———. 1969. *Canton Under Communism: Programs and Politics in a Provincial Capital, 1949–1968*. Cambridge, MA: Harvard University Press.

Walder, Andrew. 1986. *Communist Neo-Traditionalism: Work and Authority in Chinese Industry*. Berkeley: University of California Press.

———. 1995. "Career Mobility and the Communist Political Order." *American Sociological Review* 60:309–28.

———. 2002. "Beijing Red Guard Factionalism: Social Interpretations Reconsidered." *Journal of Asian Studies* 61:437–71.

———. 2004. "The Party Elite and China's Trajectory of Change" *China: An International Journal* 2 (2): 189–209.

———. 2006. "Factional Conflict at Beijing University, 1966–1968." *China Quarterly* 188 (1): 1023–47.

Walder, Andrew, Bobai Li, and Donald Treiman. 2000. "Politics and Life Chances in a State Socialist Regime: Dual Career Paths into the Urban Chinese Elite, 1949–1996." *American Sociological Review* 65:191–209.

Wan Bangru, ed. 1987. *Qinghua Daxue Fushu Zhongxue jianshi* (A brief history of Tsinghua University Attached Middle School). Beijing: Beijing Municipal Education Research Institute.

Wan Runnan. 2006. "Wo de xuezhang Hu Jintao" (My schoolmate Hu Jintao). *Qinghua Suiyue* (Tsinghua years) (April 26). http://508208.com/blog8/.

Wang Fan. 2001. "Shoufangzhe: Fan Shuo, Zhongguo Junshi Kexue Yuan yanjiu yuan, Ye Jianying zhuanji zu zuzhang" (Interview with Fan Shuo, Chinese Military Science Institute researcher and biographer of Ye Jianying). http://www.shuwu.com/ar/chinese/113606.

Wang, Shaoguang. 1995. *Failure of Charisma: The Cultural Revolution in Wuhan*. Oxford: Oxford University Press.

———. 2006a. "Regulating Death at Coalmines: Changing Mode of Governance in China." *Journal of Contemporary China* 15 (46): 1–30.

———. 2006b. "Openness and Inequality: The Case of China." In *China's Deep Reform: Domestic Politics in Transition*, edited by Lowell Dittmer and Guoli Liu, 251–82. Lanham, MD: Rowman and Littlefield.

Wang Youqin. 1996. "1966: Xuesheng da laoshi de geming" (A revolution in which students beat teachers). In *Wenhua Da Geming: Shishi yu yanjiu* (Cultural Revolution: Facts and analysis), edited by Liu Qingfeng, 17–38. Hong Kong: Chinese University Press.

Weber, Max. 1978. *Economy and Society*. Berkeley: University of California Press.

White, Gordon. 1976. "The Politics of Class and Class Origin: The Case of the Cultural Revolution." *Contemporary China Papers* 9. Canberra: Australian National University.

———. 1981. "Higher Education and Social Redistribution in a Socialist Society: The Chinese Case." *World Development* 9:149–66.

White, Lynn. 1984. "Bourgeois Radicalism in the 'New Class' of Shanghai." In *Class and Social Stratification in Post-Revolutionary China*, edited by James Watson, 142–74. Cambridge: Cambridge University Press.

Whyte, Martin. 1974a. *Small Groups and Political Rituals in China*. Berkeley: University of California Press.

———. 1974b. "Iron Law Versus Mass Democracy: Weber, Michels, and the Maoist Vision." In *The Logic of "Maoism": Critiques and Explication*, edited by James Hsiung, 37–61. New York: Praeger.

Wright, Eric Olin. 1994. *Interrogating Inequality: Essays on Class Analysis, Socialism and Marxism*. London: Verso Books.

Wu De. 2004. *Wu De koushu: Shinian fengyu jishi, yiwo zai Beijing gongzuo de yixie jingshi* (Recollections of Wu De: Chronicle of ten years of storms, some of my experiences while working in Beijing). Beijing: Contemporary China Press.

Xinhua News Agency. 2008. "Beijing University Strikes It Rich as 'Cradle of Tycoons.'" (Jan. 9). http://www.news.xinhuanet.com/english.

Xu Youyu. 1996. "Wenge zhong Hongweibing de paibie douzheng" (Cultural Revolution Red Guard factional struggle). *Zhongguo Yanjiu* (China Studies) 2.

Yan Hua. 2005. "Wu Bangguo renhu buren xiang" (Wu Banguo identifies with

Shanghai, rather than his hometown). *Yajiu shibao* (Asia times) (June 21). http://www.atchinese.com/index.php?option=com_content&task=view&id=2119&Itemid=66.

Yan, Jiaqi, and Gao Gao. 1996. *Turbulent Decade: A History of the Cultural Revolution.* Translated by D. W. Y. Kwok. Honolulu: University of Hawaii Press.

Yin Hongbiao. 1997a. "Wenhua Da Geming zhong de shehui maodun" (Social contradictions in the Cultural Revolution). *Zhongguo dangshi yanjiu* (Chinese party history) 2:77–81.

———. 1997b. "Kangzhengzhe de chongtu: Yu Luoke yu Liandong de lunzheng" (Clash of the protesters: The debate between Yu Luoke and United Action). *Zhongguo qingnian yanjiu* (Chinese youth research) 5:30–33.

Yuan Zhenguo. 1999. *Lun Zhongguo jiaoyu zhengce de zhuanbian: Dui woguo zhongdian zhongxue pingdeng yu xiaoyi de ge'an yanjiu* (On changes in Chinese education policy: A case study on the relationship between equality and effectiveness in key middle schools). Guangdong: Guangdong Education Press.

Yue, Daiyun, with Carolyn Wakeman. 1985. *To the Storm: The Odyssey of a Revolutionary Chinese Woman.* Berkeley: University of California Press.

Yusuf, Shahid, Kaoru Nabeshima, and Dwight Perkins. 2006. *Under New Ownership: Privatizing China's State-Owned Enterprises.* Stanford, CA: Stanford University Press.

Zang, Xiaowei. 1993. "The Fourteenth Central Committee of the CCP: Technocracy or Political Technocracy?" *Asian Survey* 33 (8): 787–803.

———. 2001. "Educational Credentials, Elite Dualism, and Elite Stratification in China." *Sociological Perspectives* 44 (2): 189–205.

———. 2004. *Elite Dualism and Leadership Selection in China.* London: Routledge Curzon.

Zhang Libo. 2000. "'San jie qifa' renxin han—pilin pikong yundong shimo" (Callously "letting three arrows fly"—The whole story of the movement to Criticize Lin Biao and Confucius). http://www.guxiang.com/lishi/others/youpiao/jianjie/pilinpikong1.

Zhang, Junsen, Yaohui Zhao, Albert Park, and Xiaoqing Song. 2005. "Economic Returns to Schooling in Urban China, 1988 to 2001." *Journal of Comparative Economics* 33 (4): 730–52.

Zheng, Shiping. 1997. *Party vs. State in Post-1949 China: The Institutional Dilemma.* Cambridge: Cambridge University Press.

Zheng Qian. 1999. *Bei geming de jiaoyu: Wenhua Da Geming zhong de jiaoyu geming* (Revolutionized education: The education revolution within the Great Cultural Revolution). Beijing: China Youth Press.

Zheng, Xiaowei. 2006. "Passion, Reflection, and Survival: Political Choices of Red Guards at Tsinghua University, June 1966–July 1968." In *The Chinese Cultural*

Revolution as History, edited by Joseph Esherick, Paul Pickowicz, and Andrew Walder, 29–63. Stanford, CA: Stanford University Press.

Zheng Yi. 1992a. "Taowang shengya gouqile yi duan huiyi" (An exile career evokes a memory). *Jiushi niandai* (The nineties) 6:91–95.

———. 1992b. "Shenme shi Hongweibing? Shenme shi taizi dang?" (What is the Red Guard? What is the crown prince faction?). *Jiushi niandai* (The nineties) 7:88–93.

Zhong Weiguang. 1996. "Qinghua Fuzhong Hongweibing xiaozu dansheng shishi" (Historical facts about the birth of the Tsinghua Attached Middle School Red Guard small group). *Beijing zhichun* (Beijing spring) 41:6–19. http://www.boxun.com/hero/zhongwg/11_1.shtml.

Zhou Hongjun. 2000. *Qinghua Daxue Fuzhong* (Tsinghua University Attached Middle School). Beijing: Tsinghua University Attached Middle School.

Zhou Quanhua. 1999. *"Wenhua Da Geming" zhong de "jiaoyu geming"* ("Education revolution" in the "Great Cultural Revolution"). Guangzhou: Guangdong Education Press.

Zhou Quanying. 2006. *Wenhua Da Geming shi lishi de shicuo* (The Cultural Revolution was a historic misguided experiment). Hong Kong: Milky Way Press.

Zhou, Xuegang, Nancy Tuma, and Phyllis Moen. 1998. "Educational Stratification in Urban China: 1949–1994." *Sociology of Education* 71:199–222.

Zhu, Lisheng. 1997. "Proletarian Students and the Cultural Revolution in Soviet Higher Education, 1928–1932." *Russian History* 24 (3): 301–20.

———. 2000. "The Problem of the Intelligentsia and Radicalism in Higher Education Under Stalin and Mao." *Europe-Asia Studies* 52 (8): 1489–1513.

Zweig, David. 1989. *Agrarian Radicalism in China, 1968–1981*. Cambridge, MA: Harvard University Press.

Index

East-West Center Series on

CONTEMPORARY ISSUES IN ASIA AND THE PACIFIC

Reconfiguring Families in Contemporary Vietnan
Edited by Magali Barbieri and Danièle Bélanger
2009

Southeast Asia in Political Science: Theory, Region, and Qualitative Analysis
Edited by Erik Martinez Kuhonta, Dan Slater, and Tuong Vu
2008

The Fourth Circle: A Political Ecology of Sumatra's Rainforest Frontier
By John F. McCarthy
2006

Rising China and Asian Democratization: Socialization to "Global Culture" in the Political Transformations of Thailand, China, and Taiwan
By Daniel C. Lynch
2006

Japan's Dual Civil Society: Members Without Advocates
By Robert Pekkanen
2006

The Fourth Circle: A Political Economy of Sumatra's Rainforest Frontier
By John McCarthy
2006

Protest and Possibilities: Civil Society and Coalitions for Political Change in Malaysia
By Meredith Leigh Weiss
2005

Opposing Suharto: Compromise, Resistance, and Regime Change in Indonesia
By Edward Aspinall
2005

Blowback: Linguistic Nationalism, Institutional Decay, and Ethnic Conflict in Sri Lanka
By Neil DeVotta
2004

Beyond Bilateralism: U.S.-Japan Relations in the New Asia-Pacific
Edited by Ellis S. Krauss and T. J. Pempel
2004

Population Change and Economic Development in East Asia: Challenges Met, Opportunities Seized
Edited by Andrew Mason
2001

Capital, Coercion, and Crime: Bossism in the Philippines
By John T. Sidel
1999

Making Majorities: Constituting the Nation in Japan, Korea, China, Malaysia, Fiji, Turkey, and the United States
Edited by Dru C. Gladney
1998

Chiefs Today: Traditional Pacific Leadership and the Postcolonial State
Edited by Geoffrey M. White and Lamont Lindstrom
1997

Political Legitimacy in Southeast Asia: The Quest for Moral Authority
Edited by Muthiah Alagappa
1995

CPSIA information can be obtained
at www.ICGtesting.com
Printed in the USA
LVOW03s1755210218
567418LV00003B/527/P